a land of water

a land of water

RIVERS & LAKES OF NEW ZEALAND

Pamela McGeorge

Photographs by Russell McGeorge

David Bateman

Dedication
To Erle Justice (1942–2005), a life-long friend and a frequent companion who, with his wife Judy, accompanied us on many of our adventures — on foot, in kayaks, on bikes, trains, planes and 4WD.

FIRST PAGE: Mount Taranaki with Lake Mangamahoe — where New Plymouth residents enjoy picnicking, walking and birdlife, within a few minutes of their city centre — in the foreground.

SECOND PAGE: Braided rivers are a feature of the South Island of New Zealand and provide a valuable habitat for water birds. The Arawata is an iconic river of the South Island's West Coast and lent its name to legendary gold prospector William James O'Leary, celebrated in Denis Glover's *Arawata Bill* poems.

OPPOSITE: Created by volcanic activity, this stark landscape in Tongariro National Park features Mount Ruapehu, in the background, with Mount Ngauruhoe in front, and the Blue Lake on Mount Tongariro in the foreground.

NEXT PAGE: Hoping for fish, a lone fisherman and several swans try their luck on Lake Rotoiti.

Copyright © David Bateman Ltd, 2007

First published 2007 by David Bateman Ltd,
30 Tarndale Grove, Albany, Auckland, New Zealand

ISBN 978-1-86953-616-9

This book is copyright. Except for the purpose of fair review, no part may be stored or transmitted in any form or by any means, electronic or mechanical, including recording or storage in any information retrieval systems, without permission in writing from the publisher.

Editor Jeanette Cook
Book design Think Red, Auckland
Maps Black Ant/Tim Nolan
Printed in China through Colorcraft Ltd, Hong Kong

INTRODUCTION	8
NORTH ISLAND AND SOUTH ISLAND MAPS	16
WATERWAYS OF THE NORTH Far North, Rivers of the Hauraki Plain, Waikato — River of Many Modes	18
BORN OF VOLCANOES Bay of Plenty's Rivers and Lakes, Taupo — Sleeping Volcano, Ruapehu — Crucible of Disaster	36
FROM CAPE TO CAPE East Coast Waterways, Waterways of Wairarapa and Wellington	56
RIVERS OF THE WEST Taranaki, Whanganui — Ancient River Road, Flood Waters from the Central North Island	64
NELSON LAKES AND THE WEST COAST Nelson Lakes, The Buller and its Tributaries, West Coast Rivers and Lakes	80
RIVERS OF THE EAST COAST Rivers of North Canterbury and Marlborough, Canterbury's Braided Rivers	94
BORN OF GLACIERS Otago's Legendary Lakes and Mountain Rivers, The Mighty Clutha, Taieri River and Its Catchment	110
THE FAR SOUTH Lakes and Rivers of the South	132
Index	142
Bibliography	144

INTRODUCTION

As New Zealanders, we have a natural affinity with water – with the sea and equally with our lakes, our rivers and our streams. We regard fishing as a natural right; we view swimming as a life-skill; until recently we would drink unhesitatingly from any back-country stream. For so long we, as a nation, have taken for granted the benefice of clean water in unlimited quantities.

Watched over by Mount Cook, the waters of Lake Pukaki in the Mackenzie Country change from turquoise to sky blue to dark navy when a storm is brewing.

At a symposium on the health of the Rotorua lakes, Barry Carbon, former Chief Executive for the Ministry for Environment and an Australian, told how he saw our relationship to clean water: 'It took me a while to understand that clean, fresh water is more than a value system really, it is something that flows through the veins of all Kiwis – it is part of Kiwiville.'

One of my earliest memories is of a river. It is the Manuherikia, a shallow, gravelly river that flows into the Clutha at Alexandra. As a small child I used to sit on an outcropping of schist and watch the water run by beneath me. In the hot Central Otago summers, we swam there. As an older child I spent hours building dams across streams. Later, it became an article of faith to swim every day in snow-fed Lake Wakatipu during summer holidays in Queenstown, and in winter my greatest joy was to ice-skate on the frozen dams left over from the gold-mining days around Naseby, in the Maniototo.

Shortly after I married, we moved to Waipori, a tiny village that serviced one of the earliest hydroelectric stations in the country, deep in the hills above the Taieri Plain. We lived in a dilapidated old house above the river and looked across bush to a waterfall. During holidays, we tramped alongside rivers in Fiordland and went jet-boating on rivers in Mount Aspiring National Park. I pined for these rivers when we lived overseas. As it is with most New Zealanders, rivers and lakes are an integral part of my life.

This book is about the rivers of the land and the lakes that give birth to many of them. Not all are mentioned. With hundreds of lakes, at least 70 major rivers and thousands of streams, that would be impossible. The waterways that feature have been chosen because each one has a story to tell. Like those of any community the strands are interwoven and often complicated in their unfolding.

Many rivers follow a long convoluted route, the

The Nevis River flows through some of the loneliest, most rugged areas of Central Otago.

INTRODUCTION 9

Where does the water end? Reflections in Lake Fergus, en route to Milford Sound, Fiordland.

source sometimes very distant from the area where the river is significant. The Whanganui in the central North Island is a good example. Pinpointing the location of rivers can be puzzling. The same names are often repeated and some names are confusingly similar. There are two rivers called Awatere: one in Marlborough, the other in the East Cape region of the North Island. There are six rivers called Wairoa – five in the North Island and one in the South; there are four called Waiau – two in the South Island and two in the North. The similarly named Wairau flows through Marlborough; the Waihou traverses the Hauraki Plain, and another Waihou pierces the Hokianga. The mouth of the Whanganui River is on the west coast of the North Island; the mouth of the Wanganui is on the west coast of the South Island.

There are so many lakes that some appear only as an unnamed blue shape on a map. One, deep in Mount Aspiring National Park, has the unlikely name of Lake Unknown. Two lakes each share the names of Rotoiti and Taharoa, while no less than five are called Rotoroa. In the North Island, the Rotorua area alone contains over 30 lakes, while in the glaciated areas of the South Island, lakes and ponds without number punctuate the landscape. Try attaching a map of New Zealand to the ceiling and look up at it from a prone position – a dentist's chair works well! Viewed from this perspective, the West Coast of the South Island looks ready to fray out into the sea. Fiords penetrate far inland in thick bands of blue that continue like dotted lines in a succession of lakes, big and small, right up the spine of the island. Some exist close to sea level; others perch near mountain summits.

The waterways were born so long ago that the human mind cannot grasp the enormity of the changes the story encompasses. We have to decipher their legends from the shape of the land and the residue they leave behind. Earthquakes and landslides alter the course of rivers. They in turn carry sediment, gravel and boulders that tear at the terrain they flow through. A lake formed after a glacier has retreated is a different shape from one created by volcanic action. Earthquakes and landslides also make lakes, of which the best-known is Lake Waikaremoana. Others have been less permanent. Several temporary lakes were caused by the 1929 Buller earthquake – Lake Perrine on the Mokihinui River, and Mud Lake on the Matakitaki River. Similarly, Lake Ngatapa was formed on a tributary of the Mohaka River by the 1931

A LAND OF WATER

Hawke's Bay earthquake and washed out in a flood in 1938. Lake Taupo, the largest lake in the country, covers an area of more than 600 square kilometres and was formed by volcanic eruptions.

As always, the Maori story is more picturesque than scientific explanation. Maori legends tell how the principal lakes of Te Wai Pounamu – the South Island – were created by the chief Rakaihautu. Captain of the canoe *Uruao*, which he beached at Nelson, he then headed south by an inland route. On his way south Rakaihautu used his famous ko or digging tool to dig out the biggest of the South Island lakes.

Water itself is abrasive and wears down the land. Above Lake Dunstan in Central Otago, distinct terraces mark ancient levels of the Clutha River. On the Ben Ohau Range edging Lake Ohau in Waitaki, three separate hillsides, fretted by streams, show three varying levels of erosion caused by water coursing down their flanks over thousands of years.

When people came to these shores the stories evolved in new ways. We don't know how the early Maori arrivals reacted to the profusion of rivers they found, nor how they coped with their propensity to change with little warning. But rivers, lakes and wetlands served as a larder for the earliest settlers. Freshwater fish added variety to their diet. Numerous native species existed, although some later became extinct because of competition from introduced species. Eels in particular are an important part of Maori lore. Different kinds of eel traps existed, each with a specific name, some of which are still in use today. The long-finned eel is New Zealand's biggest fish and it is estimated it has been in our rivers for at least 80 million years. In autumn, adult eels set out on an incredible journey. Starting from their home stream, they find their way down a network of rivers until they arrive at the sea and swim 6500 kilometres north into the Pacific Ocean to breed.

The tiny larvae return to New Zealand, swept along on ocean currents. Near the coast they develop into tiny transparent 'glass' eels about 60 millimetres long and head for estuaries, where they grow into elvers. In late summer, they set out upstream, often negotiating waterfalls and rapids, to return to the rivers of their ancestors.

Rivers were also an important means of voyaging for Maori, but with no wheeled vehicles, no industries and more interest in establishing their dominion over other tribes than over the land, Maori lived largely within the confines of nature as they found it, though they resorted to rampant burning in their eternal search for food.

Europeans were more inclined to see the rivers as obstacles and, where possible, as resources to harness, though they were ill-prepared for the impact that rivers would make on their daily lives. Appalling numbers of pioneers drowned while trying to make river crossings, and the first bridges they built were frequently swept away by unimagined floods. One such happened in Otago in 1878, during which, it is said, a settler and his wife found themselves floating down the river on the roof of their uprooted house. They were swept past Balclutha, right down to the river mouth where, fortunately, the house snagged on an old barn. A ship's carpenter working nearby quickly abandoned his tools, jumped into his whaleboat and rescued them before the house broke free and drifted out to sea.

Early European settlers may have arrived with little understanding of the natural forces they would have to deal with but they brought with them a determination to wrench the ways of nature into their vision of how the country should be. They built towns beside the rivers, heedless of the risk of flooding and the legacy of damage that would endure. They destroyed the forest cover on the hills without realising the effect this would have on the land and the waterways. They built bridges, drained swamps, dammed rivers and changed the shape of the lakes. And in the process they learned to be innovative. Our bridge builders and civil engineers are recognised as being among

ABOVE: Once a common means of crossing Otago's major rivers, a current-propelled vehicular ferry still operates at Tuapeka Mouth on the Clutha River. The author is contemplating the power of the current.

OPPOSITE: One of the many beautiful waterfalls on the slopes of Mount Taranaki.

INTRODUCTION 13

Willows and lupins might look pretty here, on the Cardrona River, but many of the South Island's river habitats are threatened by these and other invasive plant species.

the best in the world and are employed on numerous international projects. The development of the jet boat was inspired by the need to negotiate the shallow rivers of the Mackenzie Country, and New Zealand rowers consistently achieve success at the Olympics.

Rivers insinuate their way into our psyche and excite strong emotions. Some of the most vociferous advocates of protecting rivers from further development are anglers. Fly fishing has been an important leisure activity for many thousands of New Zealanders since trout and salmon were introduced into our rivers and lakes in the latter decades of the 19th century. Not only is New Zealand renowned for its magnificent fishing rivers, which provide opportunities for catching trophy trout, but they are also set among spectacular scenery. More recently, other recreational activities have contributed to the fame of our rivers. Overseas rafters and kayakers come to New Zealand to try their skills in both the North and South Islands. The New Zealand Recreational Canoeing Association, an active group that promotes a variety of kayaking activities, has lobbied for Water Conservation Orders and has helped to ensure intermittent releases of water on rivers where the flow is affected by hydro development.

For some activists, it is the very nature of the rivers and their historic associations that arouse their passion. The names of Titi Tihu in the Whanganui and Paul Powell in Central Otago will linger in the minds of local people for the determination with which they fought for their own special rivers – with mixed results. More recently, in the Waitaki, the combined energies of fishers, conservationists and farmers have (successfully, for the moment) fought the power of big business to save their river from further exploitation for hydroelectricity – though not for irrigation.

The impassioned national debate over the Waitaki River is a measure of how seriously New Zealanders regard their rivers. In the past cheap hydroelectricity has been a benefit from which we have all profited – committed greenie or convinced capitalist. But irrevocable changes to the landscape over the last few decades are starting to impinge on the national consciousness. There is growing concern about continuing proposals to dam more rivers, and over endless claims for irrigation. Massive intervention in what was a self-sustaining – if sometimes tumultuous – system, honed over millennia, is now seen to be irreversible and we recognise that our water supply is not infinite.

Not only the nature of the rivers is threatened. The quality of the water in many of our lakes and rivers has seriously deteriorated. Intensive land use for agriculture is causing major nitrogen pollution of waterways, as is leaching from urban areas. Some lowland waterways are so contaminated that people cannot now safely swim in them, let alone drink from them – a far cry from the days when a dead sheep upstream was the only impediment to drinking stream water.

In addition, several of our lakes are affected to varying degrees by algal blooms that can cause problems for humans as well as the interwoven strands of wildlife that the lakes sustain. These blooms are caused by the accelerated growth of algae, which produces a dense, visible patch on the surface of water. Algal blooms are generally caused by high nutrient levels, especially of nitrates and phosphorus, and are nurtured by warm temperatures. As the algae decay and die a gas is emitted that smells like rotten eggs and it can result in a total depletion of oxygen in the water. When this happens the waterway will not support life.

Fish and Game, along with other environmental and outdoor recreational groups, is actively promoting the value of our rivers and lakes, as well as smarter, less wasteful, use of our invaluable water resources. Many community groups are also involved in their local areas. In Rotorua, where several of the lakes are at crisis point, New Zealanders and overseas experts are working together on an action plan to try and control the outbreaks of algal bloom and address the conditions that cause them.

For so long, we have regarded our access to limitless clean water as a birthright. Right now this birthright is under serious threat. Rivers still

… swell and twist
Like a torturer's fist
Where the maidenhair
Falls of the waterfall
Sail through the air

as Denis Glover wrote in his sequence of 'Arawata Bill' poems, but we need to treat them with great care in the years ahead.

Tracking down rivers we didn't know, to include them in this book, turned into an odyssey of narrow gravelled roads and dead ends. Looking for the confluence of the Whanganui and the Whakapapa, we found ourselves on a track that finally became a narrow tunnel of vegetation with a mere glimmer of light at the end – and happily space to turn around – though we still haven't seen the meeting place of the two rivers.

We became adept at scrambling up and down steep banks and creeping along narrow ledges on bridges while keeping a wary eye on the traffic – and the legs of the tripod protruding onto the roadway.

We talked to a whitebaiter on the banks of the Clutha and came away with an offering of the delectable fish. We gloried in the silence of the West Coast bush as we tramped along rivers there and we shared the excitement of model yacht enthusiasts at Victoria Lake in the middle of Hagley Park in Christchurch. And the photographer tested his nerves crawling through the under-structure of the Mohaka road bridge, not far from Napier.

Many of the waterways are long-time favourites that we've enjoyed visiting in all seasons. A mid-winter tramp into the Routeburn was a special experience. Hoar frost turned the grass by the river into clumps of sparklers, lit by the sun, and the falls above the Routeburn Hut were transformed into a wonderland of ice. We have never-to-be-forgotten memories of reflections on Lake Matheson, a kayaking trip on Lake Moeraki, a mystical early morning on the shores of Lake Tutira and autumn days beside Lake Wanaka.

The waterways of this country are a gift beyond value. We must treasure them forever.

TOP: Model yachts on Lake Victoria, Hagley Park, Christchurch.

ABOVE: Lake Rotoroa, part of Hamilton Lake Domain, Hamilton.

INTRODUCTION   15

WATERWAYS OF THE NORTH

FAR NORTH

Lake Taharoa, one of the Kai Iwi Lakes north of Dargaville.

The northernmost area of New Zealand, like the rest of the country, is crisscrossed with wandering blue lines on the map indicating waterways. In this narrow stretch of land, however, many are designated as mere streams, for there is little elevation and limited space to allow significant rivers to develop. Where the term 'river' does feature it is frequently an estuary that shrinks gradually to assume the aspect of a river. Such is the Awaroa River, which meanders into the Whangape Harbour on the west coast; and both the Waihou River and the Mangamuka, which join the Hokianga Harbour, originate in the same way. The convoluted shoreline of the much larger Kaipara Harbour is similarly shaped by numerous estuarine rivers, the biggest and longest of which is the Wairoa.

It is easy to imagine the numerous canoes that must have plied these waterways when Maori were the only inhabitants of the Far North. Attracted by the equable climate and the proximity of water transport, they established sizeable communities in the area. And it was in the Far North also, that Europeans established their first missionary post in 1819 – effectively their first town – on the banks of the Kerikeri River. The famous Stone Store, at the head of the Kerikeri Estuary, is the oldest European building in the country.

Situated north of Dargaville, the three lakes – Waikare, Taharoa and Kaiiwi – are part of the Taharoa Domain. The Kai Iwi Lakes are an example of dune lakes, formed by the accumulation of rainwater in basins of sand which sit on top of relatively impermeable ironstone. A walking track follows around the perimeter of the lakes and it is also possible to walk out to the Tasman Sea, two and a half kilometres away, along a track which passes through a farm adjoining the domain.

Wairoa

The name Wairoa means 'long water' and is an appropriate name for this river, which is one of the longest navigable rivers in the country. It starts life in the middle of the North Auckland Peninsula, where the Wairua and Mangakahia rivers meet, and follows a tortuous course before it curves around the edge of Dargaville and flows in a wide, tidal channel out to the Kaipara Harbour.

For some distance from the Kaipara Heads, the eastern side of the river is bordered by extensive swamps, many of which have been reclaimed and now provide highly productive dairy land. To the west are dunes, encompassing a number of small lakes and drained swamps fringed by mangroves.

First flax, and later kauri logging, were associated with the Wairoa. Kauri gum also became an important industry. Ship building developed alongside these industries with shipyards established at Omara about 1840, Aratapu about 1880, and Te Kopuru in 1901, but when the kauri trade ceased, so too did the importance of river transport.

At one time the river and the Kaipara Harbour to the south were New Zealand's busiest waterways. However, the harbour and river are now only used for scenic cruises and recreational boating. Two of its tributaries, the Wairua, which rises in hills west of Hikurangi and joins the tidal Wairoa River near Tangiteroria, and the Mangakahia, provide good kayaking water.

Wairoa River Bridge at Dargaville.

Puhoi

It's a slow river, the Puhoi. Tidal for about nine kilometres, it winds its way in lazy loops through mangrove swamps and dense bush to the Hauraki Gulf between Whangaparaoa and Warkworth. It must have been a gloomy sight to the small group of Bohemian settlers who stepped ashore at Puhoi from Maori canoes on a wet winter afternoon in 1863. They faced a daunting task, hacking clearings out of damp

The Puhoi River loops down to its outlet at Wenderholm.

WATERWAYS OF THE NORTH

ABOVE: Bilge keel yachts are ideal in tidal mud flats such as in the Puhoi River at Wenderholm.

OPPOSITE The Waihou River wanders across the Hauraki Plain.

bush to build primitive homes and setting out to fell the timber to earn a living.

Nowadays, travellers on the Puhoi are usually there for a leisurely afternoon's paddle. If you listen carefully as you canoe your way from the village of Puhoi to the estuary at Wenderholm, you may be lucky enough to hear the mocking laughter of kookaburras. You will surely see herons, pukeko, kingfishers and other water-minded birds flying in and out of the trees along the river's path.

Tamaki

Strictly speaking, the so-called Tamaki River, which divides the Auckland suburbs of Pakuranga and Panmure, is also an estuary, extending inland like a many-armed river for 15 kilometres from its mouth in the Hauraki Gulf. On the eastern shore of the Otahuhu Creek, one of its 'tributaries', it is less than two kilometres to the waters of the Manukau Harbour, an arm of the Tasman Sea.

In the early days of European settlement the 'river' created a formidable obstruction for settlers in Pakuranga who wanted to go to Auckland. Before a bridge was constructed at Panmure in 1866, they were dependent on a punt operated by one of the 'Fencibles', assisted settlers who lived in Howick.

Crops grown in the Otara area were transported to Auckland markets in scows, shallow draft vessels which were able to negotiate the 'river' and carry on to Auckland by the harbour. Today, travellers on the highway into Auckland are more likely to view luxury launches anchored in the river. In pre-European times, Maori used the Tamaki as a highway between east and west coasts, portaging their canoes over the narrow stretch of land where the river ends.

RIVERS OF THE HAURAKI PLAIN

People travelling today in the area traversed by the Waihou and Piako rivers, see a smooth green quilt of irrigated paddocks crisscrossed with paved roads. Exotic trees grow in small clumps and neat canals channel the rivers. In 1769, when Captain Cook anchored in the Firth of Thames and rowed up the Waihou seeking wood to renew his ships' spars, the appearance of the landscape was very different. He saw a vast peat swamp, an area of extensive wetlands studded with clumps of flax and dense bush, dominated by tall kahikatea. Joseph Banks, a botanist travelling with Cook, described what they saw: 'The banks of the river were completely cloathed with the finest timber my eyes ever beheld of a tree we had seen before … thick woods of it everywhere, every tree as straight as a pencil and of immense size.'

The sailors saw water birds fossicking, where now animals graze. They took note of the occasional pa and small kainga (unfortified villages) where Maori followed the traditional lifestyle established by their pioneer ancestors. These first inhabitants settled close to the streams, using flax to make household items and fishing for eel, inanga (whitebait) in season, flounder and koura (freshwater crayfish). When they travelled, the river acted as a canoe road. Then a new wave of pioneers arrived. They looked at the country with expansionist eyes. They saw the trees and thought timber. They craved land to raise animals and crops. They needed unobstructed rivers for transport. And they discovered gold. So they logged the trees, mined the earth, drained the land, modified the rivers and changed the character of the landscape forever.

It took aeons of time and considerable volcanic action for the shallow inland sea, which once spread widely across this area, to be replaced by the low-lying land that exists today. It took less than one hundred years for it to be transformed from swamp to farmland.

The Waihou – named Thames by Captain Cook, but changed back to Waihou in the early years of the 20th century – and the Piako are the dominant rivers on the plain. Approximately 128 kilometres long, the Waihou rises mainly in the Mamaku and Patetere plateaux between Putaruru and Rotorua. For most of its length it flows north and gently eastwards, finally joining the sea near Thames on the eastern side of the Firth of Thames. It's a wide tidal river near its mouth and salt water extends up the Waihou as far as the Hikutaia River. The Piako is smaller and follows a similar path further to the west, also flowing into the Firth of Thames. Its largest tributary, the Waitoa, is born south of Matamata. These rivers drain the western slopes of Coromandel and the Kaimai Range that divides the Hauraki Plain from the Bay of Plenty.

As did the Maori, the newcomers gradually established settlements beside the waterways. Towns followed in the wake of the missionaries: Te Aroha and Paeroa on the Waihou; Morrinsville and Ngatea on the Piako.

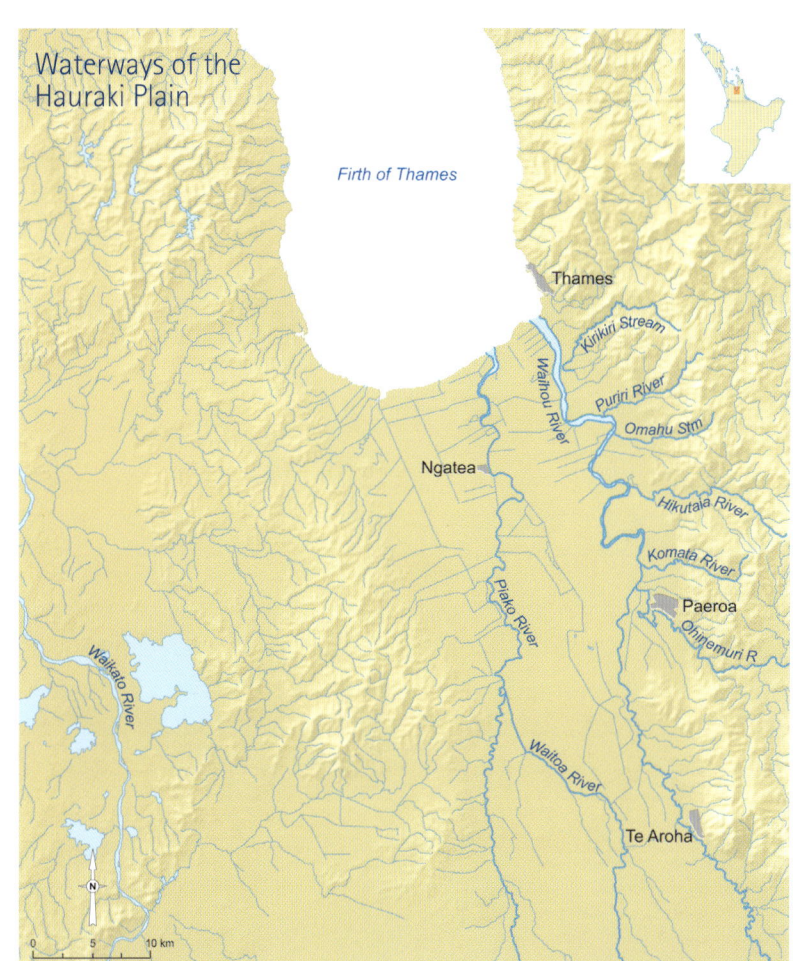

TOP: A replica of the original lifting bridge over the Piako River at Ngatea.

ABOVE: The bridge at Te Aroha over the Waihou River is a reminder of the past.

OPPOSITE: River rocks in the Ohinemuri River in the Karangahake Gorge.

River traffic

Without wheeled vehicles, Maori saw rivers as a conveyor belt – a means to move them along in the direction chosen by the water. Europeans were more impatient. They wanted to mastermind the flow and transport cargo as well. By 1878, when Te Aroha was becoming a recognised settlement, travellers were able to cross the Waihou by ferry, albeit a rather unusual one. The enterprising lessee of the Hot Springs Hotel strung a wire rope across the river and, attaching a large Maori canoe to it, ran a ferry service backwards and forwards. A few years later, Mokena, a Maori chief who had converted to Christianity, donated land around the hot soda-springs in the town and Te Aroha expanded. It gained the reputation of an Edwardian spa town, and a punt replaced the primitive ferry. An old photo shows a wagon and several wary horses being taken across the river. In 1886 a road-rail bridge came into use, with a swing section to let boats up the river. The present road bridge was completed in 1928, and the former railway bridge, not far downstream, has been converted to pedestrian use. Long-time residents still remember a stormy night in 1978 when part of a goods train was blown off the bridge. The driver of the northbound goods train stopped to clear fallen overhead wires and found 13 of his wagons had disappeared over the edge.

Paeroa came into existence as a river port on the Waihou-Ohinemuri river system as a result of the nearby Ohinemuri gold rush in 1875. By 1881 a regular passenger and cargo paddle-steamer service had been established between Auckland and Paeroa via Thames and the shipping service continued until 1947, when competition from road and railway made it uneconomic.

The passage of steamships on the rivers became possible because of the energy and determination of Josiah Clifton Firth, an early landowner in the area, who realised that his prosperity depended on river transport. Paying with his own money, he organised the removal of obstacles along the Waihou River. 'Obstacles' included large rocks, sandbanks, rapids and tree stumps: anything that might impede the progress of a ship was dynamited out of the water. Trouble occurred early, with Maori objecting to the destruction of their traditional fisheries, but Firth obstinately persisted. When he started his operations on the river, it was open to navigation for less than 50 kilometres from the mouth. By the time he had finished in 1880, shallow-draught steamers could travel at least 20 kilometres further.

Not only produce was carried on the river. Riverboat trips became a popular excursion at the end of the 19th century. Their popularity increased in the first years of the 20th century, when the district of Ohinemuri was declared 'dry'. Once a ship was moving it was deemed to be no longer on land, and alcohol could be served on board. Passengers could enjoy a tipple as they travelled to Hikutaia, where they disembarked at the hotel to continue their evening's entertainment with no traffic officers lurking on the way home!

Flood control

Floods were a recurring problem. After massive flooding in 1909, one observer described an inland sea that persisted for months, extending from above the junction of the Waitoa and Piako rivers to an area south of Ngatea. Under the 1908 Hauraki Plains Drainage Act, remedial work was started. Enormous areas of swamp were drained, rivers were dredged, canals dug and roads constructed. Willows and other obstructions were removed from the rivers and the first of many stopbanks was built.

But the flooding did not abate, and during the 1950s and early 1960s storms caused havoc. As a result, two ambitious flood control projects aimed at protecting a large part of the Hauraki Plain were conceived and constructed. The first, called the Piako River Protection Scheme, was built between 1962 and 1979. It starts at the foreshore near Thames and, in all, consists of a series of floodgates, pumping stations and extensive stopbanks. Maintenance is ongoing.

The Waihou Valley Scheme, conceived in the 1960s with construction running from the 1970s into the 1990s, provides flood protection, river management, land drainage and soil conservation in the Waihou River Valley and is the largest flood control scheme in New Zealand.

The Ohinemuri River

The rivers of the Hauraki Plain are smooth and slow-moving as befits lowland rivers. But the Ohinemuri, child of the Waihou, rises in the ranges behind Whangamata and rushes through the rugged Karangahake Gorge to meet its parent at Paeroa, and is constrained by rocky cliffs and broken frequently by rapids. In flood it becomes a muddy, raging monster, as residents of the area learned in 1981 when 10,000 hectares of the Hauraki Plain were flooded and the small town of Waikino in the Karangahake Gorge was demolished. The Ohinemuri rose by several metres, and I can remember travelling beside the river months later and seeing sheets of roofing iron wedged in trees high above the roadway.

Before European settlers arrived, the stony bottom of the Ohinemuri River was home to thriving populations of whitebait, eel and other native fish. But all this changed when the Ohinemuri Goldfield opened in 1875. The settlements of Mackaytown,

---

ACTIVITIES

FISHING THE WAIHOU RIVER SYSTEM:

WAIHOU RIVER: Most fishing is carried out in the upper reaches above Okoroire, where the Waihou is very clear – classic dry fly waters. In this section of the river, trout numbers are very high.

WAIOMOU STREAM: An upper tributary of the Waihou; about 20 km of fishable water. Popular river with a high catch rate. Three tributaries of the Waiomou – the Omahine, Rapurapu, and the Kakahu streams – offer good small stream fly fishing.

OHINEMURI RIVER: Excellent fishing throughout the Ohinemuri, especially upstream of Waihi and through the Karangahake Gorge. About equal numbers of rainbow and brown trout.

WAITAWHETA RIVER: Boulder stream flowing through forest and farmland catchments before joining the Ohinemuri River. Highly valued by anglers for its scenic beauty and feelings of solitude.

WAIMAKARIRI STREAM: Large, spring-fed stream, flowing north from the Kaimai Ranges to meet the Waihou River just south of the Okoroire Falls. Renowned for its high catch rate of smallish trout. Larger trout present in the upper reaches. Probably one of the best dry fly streams in the region.

Part of the Victoria Battery for the Martha Mine — a reminder of the gold mining history of the Coromandel.

Owharoa and Waikino along the Karangahake Gorge grew rapidly to service the timber milling activities and the stamper batteries, used to extract the gold. The river was declared a 'sludge channel' in 1896 and tailings, mining debris and waste water from the quartz-crushing gold extraction plants in the Karangahake Gorge and Waihi were discharged directly into the river. Between 1896 and 1952, when the last gold and silver processing battery at Waikino closed, suspended sediment levels in the river increased hugely, to the detriment of fish life and water quality. Protest at the degradation of the river, from both Maori and Pakeha, had little influence over the years.

Once again, however, there is change. When the Victoria Battery was constructed to process ore from the Martha Mine at the end of the 19th century, its site was chosen because of the availability of relatively cheap hydroelectric power from the Ohinemuri and Waitekauri rivers. The river was seen as a resource to be exploited. The original Martha Mine closed in 1952, but when it reopened in 1987, the old pioneering attitude of expansionism had swung towards protection of the environment, if not total conservation.

The Martha Mine of today functions under an atmosphere of strict compliance to consents issued under the Resource Management Act. Producing more than one million dollars' worth of gold and silver per week, the open pit mine operates with a surplus of water. Water management is an integral part of the operations. A water treatment plant was built as a requirement of the mining licence and has been extended to allow the mine to operate for a further few years. It exists to treat surplus water before discharging it into the Ohinemuri River, ensuring that the biology of the river is protected as well as the rights of the downstream users. Not only is the water treated to remove chemicals used in the extraction process, but gauges also ensure that the amount of water discharged each day complies with the granted consents.

Alongside the Ohinemuri today there is little sign of the feverish activity that took place when the Victoria Battery was in operation. The Karangahake Gorge, with the picturesque river at its feet, has become a favourite haunt of picnickers and hikers who come to investigate the remains of the mining era, to walk along the tracks that have been constructed through the gorge and to enjoy the arts and crafts offered for sale along the route.

Owhara Falls on the upper Ohinemuri River.

A popular walkway runs alongside the Ohinemuri River, through the Karangahake Gorge.

WAIKATO — RIVER OF MANY MODES

OPPOSITE TOP: Port Waikato, mouth of New Zealand's longest river.

OPPOSITE BOTTOM: Almost 425 kilometres upstream, the Waikato River enters the shute leading to the Huka Falls, near Taupo.

To an engineer the Waikato River is a source of power; to an angler and a canoeist it's a source of sport; and to a Maori who traces his descent from the great *Tainui* canoe, the river is important as the source of life and mana. It also feeds the largest lake in New Zealand.

For five centuries before Europeans explored its length and set out to harness its energy, this powerful river provided all that was important to a people whose lives were tuned by seasonal rhythms and not by the urge to change the face of nature. Now the hydroelectric power produced from its waters provides a significant portion of the country's total power needs.

The river's course

It's a long river, the Waikato: the longest in the country. It starts life in the Tongariro National Park, gifted to the nation in 1887 by paramount chief Tukino Te Heuheu of Ngati Tuwharetoa. For 425 kilometres it wanders and winds its way north from a bubbling stream on the slopes of Mount Ruapehu, through the immensity of Lake Taupo and the turbulence of the Huka Falls, before submitting to the discipline of hydro dams and city riverbanks, and gradually sliding out to sea at Port Waikato through a delta of marshy islands and backwaters.

As it makes its way towards Lake Taupo it becomes entangled with the Tongariro River system, often considered part of the Waikato itself. A total of 47 streams and rivers feed Lake Taupo, but only one river drains it – the Waikato.

Shaped by often violent upheavals across the North Island, the river has not always followed this course. Once it continued to flow east after emerging from Lake Taupo and followed a route through present-day Reporoa and out to the Pacific in the Bay of Plenty. Later it changed again, flowing in a more northerly course to meet the sea in the Firth of Thames. Its present route probably dates from about 10,000 years ago.

On its long transforming journey, the river starts life amid snow and glaciers, flows through desolate areas of volcanic rock and arid desert, past stately beech forest and plantations of pine, steaming volcanic

A tranquil stretch of the Waikato River downstream from the Whakamaru Dam.

A LAND OF WATER

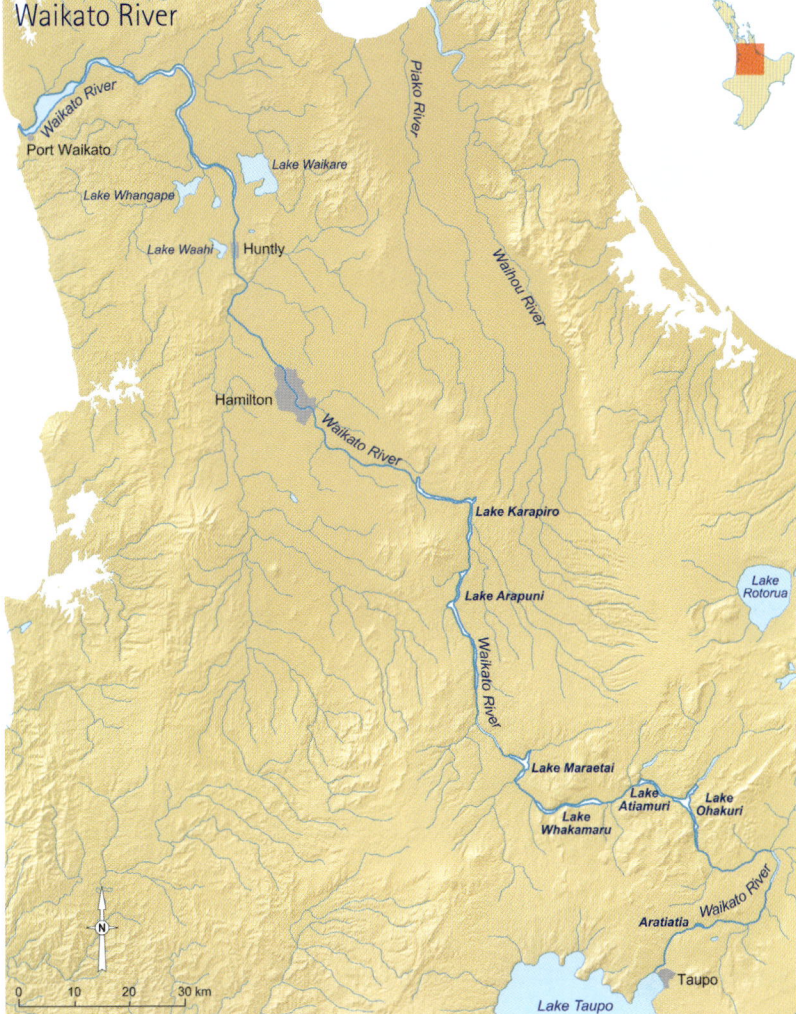

cliffs, prosperous green paddocks and eventually black, steel-producing iron-sand. It is fed by a multitude of small streams and bigger rivers. It swells to form sinuous lakes and is diverted through tunnels. It roars through the turbines of eight hydroelectric power stations, feeds on sewage from eight towns and one city, and is mixed with treated waste from 10 or more major industries. Warmed by the Huntly power station, it runs over a coalfield where miners once worked and listened to the sound of riverboats passing over their heads. Enriched by fertilisers, pumped into irrigating systems and drunk by 140,000 people, this is a river of multiple uses.

Early settlers

It is thought that people first came to the Waikato region about AD 1300. They were Maori of the Tainui tribe and they built fortified villages on the banks of the Waikato River near the area that Hamilton covers today, which they called 'Kirikiriroa'. It was a river of plenty, offering an abundance of fish – eels, whitebait, freshwater crayfish (koura), mullet and waterfowl; in the swamps along the riverbanks, wild vegetables such as watercress were harvested. It was also a river that brought conflict on its current, and war canoes were a recurring sight through the centuries.

As well as providing a network for travel and communication, the river was an integral part of life's rituals and became embedded in the Maori concept of spirituality that creates a sense of continuity from generation to generation. It is this sense of mauri, or life force, which Waikato-Tainui iwi feel has been violated by European exploitation of the river and which formed part of their claim under the Treaty of Waitangi.

ACTIVITIES

Canoeing, rafting, jet-boating, sailing, wind-surfing, water-skiing, duck-shooting and trout-fishing are all available on the region's rivers and lakes.

At Lake Karapiro, world-class rowing facilities are available.

The Waikato River trail, situated between Waitomo, Rotorua and Taupo, is planned to cover approximately 100 kilometres and will allow people to explore lesser-known areas in the region. The trail will reflect the heritage and natural features of the area and passes along the mid-Waikato River from Karapiro to Atiamuri, linking six lakes in the area.

Europeans visited the Waikato in the 1820s, but it was another 20 years before settlers arrived. First came the traders, followed by missionaries, and for almost another 20 years there was harmony between the two races. The Waikato became a centre for commerce and the river a vital link with markets in the north. Maori transported produce – flour, maize, pigs, potatoes, kumara and other vegetables, flax, timber and fish – via the river to Onehunga and Auckland. But relations soured with European expansion.

When the Waikato tribes elected their own king in 1859, it was a signal by Waikato Maori – who had not signed the Treaty of Waitangi – that they wanted to exercise responsibility over their own territory. The war that erupted in 1863 seemed to them like a hostile government response to that election. In fact it was the result of a build-up of tension between the two races and a reluctance on the part of Maori to sell more land. In July 1863 when General Cameron and his troops crossed the Mangatawhiri, his action amounted to an invasion.

The war continued for 15 months with Maori hopelessly outnumbered by British troops. Given the choice of submitting to Queen Victoria and surrendering their arms and land, the Waikato people preferred to withdraw from around the Waikato and the Waipa – in effect a stalemate, as the British did not pursue them.

But it was the aftermath of the war that caused lasting bitterness among the iwi of the Waikato. The vast government confiscation of land, conceived as punishment for any tribes who rebelled against the Crown, in fact provided land for soldiers to buy up and thus help pay for the costs of a war that was started by the government. Some tribes, such as Ngati Maniapoto, who participated in the wars, escaped the full force of the confiscation, while the central Waikato tribes lost almost all their lands. Thus were laid the foundations for ongoing poverty among the Maori of the area and more than 100 years of agitation for return of their land.

Not until May 1995 did the Crown finally sign a Deed of Settlement with Waikato-Tainui that redressed, in Maori eyes, the wrongs committed so long ago.

The river as a road

In his book *Bow Waves on the Waikato*, Graham Vercoe describes the Waikato as a highway for exploration, settlement, trade and warfare. For centuries, simple waka and more intricately decorated war canoes were the only vessels that plied the river. It is not clear when the first European ship entered Waikato waters but it may have been in 1826, when the flax trader John Kent sailed at least some of the way upriver. But Europeans initially travelled in Maori canoes, and not until the threat of war arose did the Colonial Government, or indeed any private traders, make a move to establish a regular riverboat service.

In 1863 the first paddle-steamer, *Avon*, was bought by the government and fitted out as a gunship. The *Pioneer* arrived on the river soon after the *Avon*, and it too was used during the war. Later it carried settlers and supplies to the towns of Cambridge and Hamilton, both of which were planned as garrison towns established to deter any Maori attempts to

This tourist boat on the Waikato in Hamilton is a replica of the early paddle-steamers that used to ply the river.

A LAND OF WATER

reclaim territory. Several other steamers were operated by the privately owned Waikato Steam Navigation Company. When trains eroded the steamer business in the 1880s, the company started running picnic cruises on the river and moonlight dance trips in the summer. Although the heyday of river travel was over by the beginning of the 20th century, riverboating was revived by Waikato-born Caesar Roose. He started a service carrying passengers and freight to Port Waikato in the early 1900s and his business continued through the 1930s, with river cruises proving a popular leisure pastime. River transport declined again in the 1940s, although there were barges operating for another couple of decades. Nowadays, a replica paddle-steamer called the *Waipa Delta*, powered by a jet engine and built in the 1980s, runs cruises around Hamilton.

Ornately carved canoes have also reappeared on the river. Confiscated along with the land after the 1860s war and often destroyed, war canoes seemed threatened with extinction. But recent decades have seen a resurgence in traditional Maori carving. Several ceremonial canoes have been lovingly constructed and every year at Ngaruawahia, at the location of the Turangawaewae marae, magnificent waka take to the water as part of the historic Ngaruawahia regatta. Turangawaewae marae, on the banks of the Waikato, is the official residence of the head of the Kingitanga – until her death on the 15 August 2006, this was Te Arikinui Dame Te Atairangikaahu, the sixth Maori monarch, who was succeeded by her eldest son, Tuheitia Paki.

The river today

The land that was occupied by military settlers after being confiscated from Maori had a very different appearance from the land that borders the Waikato today. Huge, reedy marshlands filled the Waikato Basin. Swamps defeated many of the original small landholders; the rich dairying land, orchards and stud farms that now spread across the Waikato are the result of much hard labour and a formidable drainage programme that only large landowners were able to afford.

Through the first half of the 20th century it was dairying that led to the growth of the city of Hamilton and its surroundings. Since then the Waikato dairy industry, nourished by the fertility bestowed on the area by silt from the river, has continued to be an important ingredient of the national economy. The growth of tourism is another factor in the Waikato's economy as recreational activities on Lake Taupo and the network of rivers and lakes attract increasing numbers of visitors.

However, the greatest contribution made to New Zealand life by the Waikato River is the electricity produced from its waters and the thermal resources close by. From 1913, when the first hydroelectric power station was built at Horahora to provide electricity for the Waihi Gold Mining Company, there has been a continuing programme of construction along the river culminating in the largest plant, the Huntly coal-fired power station, commissioned in 1983, which can provide up to 20 percent of the country's electricity needs.

TOP: Lake Maraetai Dam, one of eight on the Waikato River.

ABOVE: Indulging the need for speed – jet-skiing on Lake Karapiro.

RIGHT: A maimai on Lake Whakamaru.

The Tongariro Power Development scheme, centred in Turangi and started in the 1960s, involved collecting water from the central volcanic plateau, specifically the upper reaches of the Whanganui River, and diverting it through tunnels to Lake Taupo and the Waikato power stations. In this way, maximum storage was maintained while the outflow from the lake was increased, as was the generating capacity of the eight hydroelectric stations on the Waikato River, though it led to considerable protest from those who cherish the Whanganui and its waters. The hydro lakes on the Waikato are Lake Aratiatia, Lake Ohakuri, Lake Atiamuri, Lake Whakamaru, Lake Maraetai, Lake Waipapa, Lake Arapuni and Lake Karapiro.

The run-off from farming and horticultural activities, and the production of electricity, has caused enormous changes to water quality and river flows, and to the shape of the river. While water sports have in many cases benefited from the creation of lakes such as Karapiro and Maraetai, the number of fish species has declined.

In the 21st century there is greater emphasis on assessing the impact of industry, and future development will increasingly depend on the ability to minimise damage to the river.

Waikato peat lakes

In the environs of the Waipa and Waikato rivers are numerous small lakes and swamps created when tributary streams were stranded by the build-up of silt and volcanic ash washed down over centuries from the Taupo area. Often they have deteriorated into murky, weed-infested pools of peaty water, yet they are potentially valuable sanctuaries for wildlife. There are moves afoot to rehabilitate them. In 2005, a World Wetlands Day was held in Rangiriri, focusing on the importance of wetlands and ways of restoring those that have been degraded.

Lake Rotomanuka, near Te Awamutu, is administered by the Department of Conservation (DOC) as a wildlife reserve. Much smaller is Lake Kaituna, where significant changes have been made by Andrew Hayes, the farmer whose property surrounds the lake. Now there are fences around the metre-deep wetland, native species have replaced exotics, and feral cats and possums have been eradicated. 'We used to have the cows in the wetland feeding off it and pooping in the water, and the Canada geese and mallards fouling our pasture,' said Hayes. 'Now we have the cows in the paddocks and the birds in the water, the way it ought to be.' In recognition of their work the Hayes family received a Heritage Restoration Award at the 2005 regional farm environment awards.

Conservationists with similar concerns about the state of riverbanks and the quality of stream water in dairying country near Raglan, are encouraging farmers to fence their land to keep stock away from waterways and to plant native vegetation along riverbanks. Where this has already occurred there has been a fast and noticeable improvement in the quality of the streams.

ACTIVITIES

The Waikato River, along with its associated wetlands and lakes, is home to 19 native fish species and at least 10 introduced species. The river supports a large recreational whitebait fishery, a large commercial eel fishery and a winter trout fishery.

Brown trout are abundant throughout the Waikato River. Over summer, they live in the cool headwaters of the Waipa River, but after spawning in April/May they move downstream, remaining in the river until December.

Rainbow trout are plentiful in the Waikato River in Hamilton and upstream to the Karapiro Dam – though generally smaller than brown trout – and are present in the river all year round.

Best fishing is in the winter months, when water clarity is high due to the decline in summer algal growth.

A LAND OF WATER

The discharge from the Aratiatia Dam down the Waikato River is staged several times each day for tourists.

WATERWAYS OF THE NORTH

LEFT: The mighty Huka Falls on the Waikato River.

TOP: The Waikato River upstream of the Huka Falls, near the exclusive Huka Lodge (shown in the photo below).

WATERWAYS OF THE NORTH

ABOVE: Bueller's Landing is a popular picnic spot on the Waikato River, above the Arapuni Dam.

RIGHT: Early morning boating on Lake Maraetai.

A LAND OF WATER

The railway (near) and road (far) bridges over the Waikato River at Hamilton.

WATERWAYS OF THE NORTH 35

BORN OF VOLCANOES

BAY OF PLENTY'S RIVERS AND LAKES

A river meanders across the plains with Mount Edgecumbe in the background.

Look at a map of the Bay of Plenty and you see a wide sweeping crescent of coastline lapped by the Pacific Ocean. Not far inland is Lake Rotorua, largest in a broken chain of lakes formed at various times in the last 150,000 years by violent volcanic action. North and east of the lakes lie the remote ranges of Kaimai, Te Urewera, and Raukumara – not immensely high but sufficiently rugged to give birth to a multitude of fast-flowing streams and rivers that stitch and seam the land before emptying into the ocean through troublesome flood plains and wide estuaries that have long been a source of seafood.

These rivers were once roadways and eel fisheries for Maori, tumultuous in their headwaters, calmer near the coast. Now they are a playground for everyone who enjoys pitting their skills against the whims of currents and rapids, or their wits against those of wily trout. But these rivers also have a serious side. Some are harnessed to provide hydroelectric power, and when storms lash the Bay of Plenty, the flooding rivers are a reminder that the power of nature is still omnipotent.

The lakes, formerly stopping places for wandering Maori iwi, have become centres for tourism and holiday-makers.

Rotorua is a land of lakes and rivers, geysers, hot pools, dozing volcanoes and colourful Maori legends. Arriving in the area by car, the visitor sees a green and pleasant land, forest hiding the rough contours of past upheavals and cultivated paddocks disguising the harsh volcanic residue beneath the pasture. But occasional clouds of errant steam and the whiff of sulphur are constant hints of the unquiet earth beneath. Seen from the air, the contorted nature of the land becomes more obvious. A chain of crater lakes, hills shaped by volcanoes and steaming rivers all tell their story.

In all, 11 major lakes and numerous smaller ones make this a paradise for anglers and anyone who enjoys water sports in a warm climate – not to mention the advantages of a natural, 24-hour-a-day hot-water system that can be diverted for heating houses and enjoyed in therapeutic spas.

But this is also a landscape prone to cataclysmic

change with scant warning, as happened in 1886 when Mount Tarawera blew its top. The massive eruption wiped out the famous Pink and White Terraces, as well as destroying three Maori villages, killing around 150 people and depositing a thick layer of ash and pumice over the surrounding land.

The area was initially settled by Maori of the Te Arawa iwi, probably about the 14th century, and small villages occupied sites beside many of the lakes. European missionaries arrived in the 1840s, and it wasn't long before the extraordinary character of the landscape started to attract visitors. During the period of the New Zealand wars in the 1860s, there were several serious skirmishes around the shores of Lake Rotorua; but a mere 20 years later Rotorua was accorded special status as a spa town, establishing the earliest tourist industry in New Zealand.

Tourists also thronged to the village of Te Wairoa, which developed as a flourishing centre for Victorians wishing to visit the Pink and White Terraces. Visitors stayed at the Rotomahana Hotel in Te Wairoa, were ferried across Lake Tarawera to Te Ariki and from there walked to Lake Rotomahana where the terraces were located. Unexpected prosperity came with the tourists, but all that changed with the eruption of Mount Tarawera. The survivors, devastated by the tragedy, left the area, and many settled at Whakarewarewa, much closer to Rotorua.

Lake Rotomahana was totally emptied by the Tarawera eruption, leaving craters, steaming holes and pools of hot mud. When it refilled it formed a larger lake than before. Tourists visiting the Waimangu Valley can take a boat tour on the lake and experience at close quarters some of the geothermal activity on the water's edge. Although fewer anglers fish Lake Rotomahana than some of the other Rotorua lakes, it is said that the purest strain of rainbow trout in the world live here and they are reputed to possess excellent fighting qualities.

Rotorua's lakes are beautiful and they provide important recreational facilities for visitors and locals alike. They have long been favoured holiday destinations, and baches are scattered through the bush near their shores. Two and three generations of families have come to spend their summers swimming, boating, fishing and enjoying the luxury of hot-water springs to relax in when the long hot days are waning.

But problems have appeared in paradise. For decades the pressures caused by increases in population, escalating visitor numbers and farming have all taken their toll on the environment, particularly on the quality of the water in many of the lakes. Now toxic algal blooms regularly appear, diminishing the oxygen content of the water, causing an evil-smelling green-brown slime and killing fish.

Lake Rotorua itself has a city of about 69,000 people on its shores, as well as smaller communities scattered around its edge. A new sewerage scheme installed in the late 1980s improved the quality of the lake water temporarily but it has since deteriorated again. The main problem is the excess of nutrients – particularly nitrogen and phosphorus – that drain into its waters and those of other lakes, via streams and rivers, from the farms within their catchment areas.

Environment Bay of Plenty, Rotorua District Council, Te Arawa Maori Trust Board and dedicated community groups are working to protect and hopefully improve the water quality of several lakes. Action Plans for lakes Rotorua, Rotoiti, Okareka, Okaro and Rotoehu have been developed. Their main focus is on how to reduce nutrients already in the water as well as finding ways to cut back on the amounts of nitrogen and phosphorus leaching into the lakes. Symposia, jointly hosted by the Lakes Water Quality Society and the Royal Society of New Zealand, have been held annually since 2001 with the aim of finding solutions to the problem and instigating continuing action. Ensuring that the whole community is aware of the problem and educating everyone to take responsibility is an important part of the process.

Waiting for the day's influx of visitors and tourists – Lake Rotorua.

The Rotorua lakes

BELOW: Evening peace at Lake Rotoiti.

BOTTOM: Lake Rotoehu, one of four lakes in a chain linking Lake Rotorua with its sister lakes Rotoiti, Rotoehu and Rotoma.

The largest of the lakes, and also the oldest, is Lake Rotorua. Relatively shallow, it was formed about 140,000 years ago and its only outflow is the Ohau Channel, which flows into Lake Rotoiti and which in turn is connected by subsurface flows to Lake Rotoma and Lake Rotoehu. These last three lakes are of much more recent origin, having been formed about 8500 years ago in the Rotoma eruption. Even newer is tiny Lake Okaro, a mere baby at 800 years of age.

Lake Rotorua will be forever famous for the story of the two lovers, Tutanekai, a commoner who lived on Mokoia Island in the middle of the lake, and Hinemoa, daughter of an aristocratic family who lived on the lake shore. Forbidden to meet, Hinemoa eventually swam to her lover, and their respective families finally bestowed their blessing on their union.

The Rotorua lakes all provide excellent fishing for first-timers as well as experienced anglers. Most are readily accessible and well signposted.

With a reputation as an immensely popular lake for holiday-makers and anglers alike, Lake Rotoiti is suffering from algal bloom problems that have seriously affected its ability to sustain fish life. As with Siamese twins, whatever affects Lake Rotorua also affects Lake Rotoiti. With no large town on its shores and few farms in its catchment, the pollution derives mainly from Lake Rotorua. Around 70 percent of the lake's nutrients have come via the Ohau Channel, and there are plans to build a wall to divert the channel away from the main body of Lake Rotoiti and down the Kaituna River. It is hoped this will improve the lakes significantly and reduce algal blooms. Money is also being allocated to encourage fencing and planting along the lake shore.

Native bush and tree ferns cluster on the hills around Lake Tarawera. Here and there holiday houses peer out among the trees and boat sheds dot the shore. A popular boating lake, it also attracts anglers from around the world in search of trophy fish – rainbow trout that can reach five kilos. Holiday-makers enjoy swimming, followed by a warm-up sitting in a scooped-out hollow of the beach where natural hot water bubbles up through the gravel at the edge of the lake. And if by chance you should be on the water, and see a ghostly Maori war canoe approaching, be afraid. In 1886, shortly before Mount Tarawera blew up, numerous tourists reported seeing such a phantom canoe gliding across the lake before mysteriously disappearing in the mist.

The Lake Okataina area was an important link in pre-European routes, when canoes were carried from Lake Tarawera to Lake Okataina. Nowadays, the Okataina Road follows one of these ancient portage routes. A scenic reserve, rich in birdlife and containing specimens of rimu, totara, rata and kahikatea, extends right to the lake's edge in one area. There are thermal springs on the eastern shore and boaties launch their craft off the open beaches.

Also called the Blue Lake, Lake Tikitapu is tiny and is situated between the lakes Okareka and Rotokakahi, (known also as the Green Lake). It is a popular fishing lake. Rotokakahi was home to the Tuhourangi tribe before the Mount Tarawera eruption and is named for the shellfish Maori call kakahi, which was part of the local diet and was used especially to nourish motherless children. When cooked and softened with water, it could be sucked like milk. More than a century ago a group of Nga Puhi living on the island in the middle of the lake was massacred, and a tapu has been placed over the lake. No one lives here now, but in times gone by the lake was heavily populated and it was once a popular tourist destination. The lake is closed to non-Maori.

Looking at Lake Rotoehu on a map, it has the appearance of a stunted tree with several arms at the northern end representing branches. Although only 25 minutes by car from Rotorua, such is the atmosphere at Lake Rotoehu that it feels as if you have entered a wilderness. It provides good fishing for rainbow trout, especially from a boat.

Lake Rotoma, a popular holiday and fishing destination, is the only lake in the eastern region where tiger trout are released. It has a reputation for 'gin clear' water, but recently there have been concerns about the effects of cattle drinking in the lake, and there is a risk that future development on the shore will see seepage from septic tanks.

ACTIVITIES

Most of the lakes (excluding Rotokakahi) are open for fishing and a variety of water sports including water-skiing, kayaking, swimming and soaking in hot-water pools.

LAKE TARAWERA: Boat charters, fishing, jet-skiing.

LAKE ROTOMAHANA: Wildlife refuge. No boat ramps around the shore, but dinghies can be launched off the Waimangu loop. Rainbow trout fishery.

LAKE OKAREKA: Boat launching facilities at Acacia Bay.

LAKE ROTOEHU: Almost essential to have a boat to fish this lake successfully. No concrete ramps. Depending on lake level, launching is available at Kennedys Bay, Otautu Bay and, with care, at Te Pohue Bay.

LAKE TIKITAPU: All-weather boat ramp at the main beach; walking track around the lake making it easy to find a good fishing spot on land.

LAKE ROTOKAKAHI: Privately owned and only members of the Tuhourangi iwi allowed access onto the lake. No camping is permitted near the lake or on its island. From the bush walk around the Blue Lake, it is possible to view Lake Rotokakahi.

Ready for kayaking, Lake Okataina.

BORN OF VOLCANOES 39

ABOVE: Surrounded by bush, Lake Tarawera is a popular holiday destination.

RIGHT: Homes and boat sheds on the shores of Lake Rotoiti.

A LAND OF WATER

LEFT: Mokoia Island, Lake Rotorua, is famous for the Maori legend which tells of Princess Hinemoa who swam from the lake edge out to her lover Tutanekai, who lived on Mokoia.
BELOW LEFT: A house nestled among tree ferns on Lake Tarawera.
BELOW RIGHT: (top) Fishing on Lake Rotoma.
BELOW RIGHT: (bottom) Boats moored in the reeds on the edge of Lake Tarawera.

BORN OF VOLCANOES 41

One of the smaller lakes, Okareka sits between Lake Okataina and Lake Tarawera, linked to the latter by the Waitangi Spring.

Lake Tikitapu, also known as the Blue Lake.

OPPOSITE: The dramatic Okere Falls, at seven metres, is the highest commercially rafted waterfall in the world.

42   A LAND OF WATER

BORN OF VOLCANOES 43

White-water thrills

Although short, the Wairoa River has a reputation that exceeds its size. Running through a rock- and tree-lined gorge that combines turbulent rapids with exciting drops, it is recognised worldwide for its rafting value, and sportsmen come to experience its top white-water action. The Wairoa flows out to the Bay of Plenty near Te Puna and as well as providing great opportunities for water sports, it also supports several small hydroelectric power stations, the first of which was constructed in 1915. Today the scheme consists of the Kaimai 5 Station on a diversion tunnel feeding Lake Mangaonui, the Lloyd Mandeno station, sited on the west bank of the Mangapapa River, the Lower Mangapapa Station, and four kilometres further downstream, the Ruahihi Station.

In 1989 the McLaren Falls Station, built in the 1920s, was decommissioned, and a bypass was subsequently installed to allow the continued release of recreational flows into the Wairoa River on set days each year for activities such as rafting and canoeing.

Another river is the Kaituna River, which starts life at Lake Rotoiti and is renowned as a thrilling ride for white-water rafters. It no sooner emerges from the lake than it tumbles over the Okere Falls, a drop of seven metres, the highest commercially rafted rapid in the world. A few remnants of an early power station, built on these falls in 1901, can still be seen in the bush. Until it was decommissioned in 1939, it provided power to Rotorua. Behind and around the waterfall, ancient caves exist that once provided shelter to women and children during intertribal Maori battles. Now they are home to millions of glow-worms.

The Kaituna catchment area includes both Lake Rotoiti and Lake Rotorua, as well as several tributary streams that drain the hill country behind Te Puke. The Mangorewa, in particular, contributes greatly to flood flows in the Kaituna. After emerging from the steep, narrow gorge that makes this river so exciting for rafters, it winds its way to the sea at Maketu by alluvial terraces and through the peat and sand deposits of a large delta area that is prone to flooding. In an attempt to limit the effects of floods, most streams in the Upper Kaituna area have been fenced and planted. Stopbanks have been built through the urban areas of Rotorua and structures have been erected to control water levels on the two lakes. In the Lower Kaituna

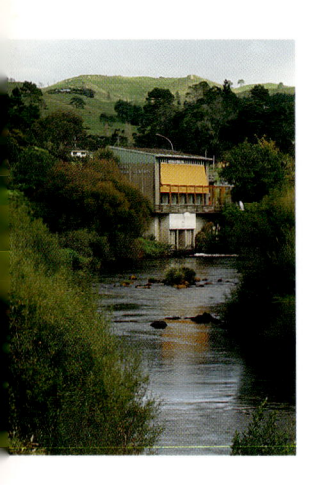

ABOVE: The Ruahihi Power Station on the Wairoa River.

RIGHT: The picturesque Kaituna River.

considerable flood protection measures have also been taken, including the building of stopbanks, canals and drains, and five major floodgates.

River of the god

The Tarawera River emerges from Lake Tarawera and three kilometres downstream from the lake it disappears underground, to resurface later as a cascade from a vertical rock face, the Tarawera Falls. Only 65 kilometres long, the river now runs where lava once poured from the eruption of Mount Tarawera. Near the coast, it is a smooth, wide ribbon of silver, flowing quietly to the sea.

Legend has it that its original name was Te Awa-a-te-Atua (river of the god), reflecting the awe in which its discoverer, Ngatoroirangi, was held. High priest on the Te Arawa canoe and ancestor of Ngati Tuwharetoa, he parted company with other members of Te Arawa after their arrival from Hawaiki, and moved down the coast with his followers, giving name to the river that flows into the sea at Matata, west of Whakatane. Later he turned inland to explore, where he caused the goddesses of fire to come to his rescue when he was dying of cold on Mount Tongariro, and thus was responsible for establishing the central North Island geothermal area over which the people of Te Arawa were recognised as guardians.

The power of the Rangitaiki

The Rangitaiki River begins life on the northeastern slopes of the Ahimanawa Range, flowing through the Kaingaroa Forest in its upper reaches, and tumbling through narrow gorges interspersed with calm stretches on the plains of the upper Rangitaiki and Galatea. Two large tributaries, the Whirinaki and the Wheao, join the main river not far from the township of Murupara, before it drains into Lake Aniwhenua, formed in 1980 by damming the Rangitaiki River for hydropower generation. The waters of the Rangitaiki also contribute to the Wheao Hydroelectric Scheme. A second dam at Matahina provides water storage for hydropower, and from there the river continues to the coast, meandering through the flood-prone Rangitaiki Plains and out to sea not far from the town of Edgecumbe, at Thornton.

Also used for recreational purposes, in the words of a canoeist it is: 'A great first-time river. Exciting, long, bouncy, white-water, larger rapids near the start of the paddle. A quiet stretch in the middle followed by fun rapids through to the finish.'

In times gone by, the Rangitaiki and its tributaries served as a vitally important food source and means of transport and communication for Maori. The construction of the dams caused a major disruption to the traditional eel fishery on the river. Now eels, migrating both upstream and downstream from Lake Aniwhenua, are transferred past the barrier, and it seems that eel stocks are being successfully maintained.

Flooding on the low-lying coastal area is an on-going problem. The Rangitaiki-Tarawera river scheme was designed to provide protection from a so-called 100-year flood, but in July 2004 a breach in the stopbank just above Edgecumbe resulted in massive flooding. This was the third major disaster to hit the Rangitaiki Plains in 17 years, following major floods in 1998 and the Edgecumbe earthquake of 1987, which also caused major damage to the stopbanks along the Rangitaiki River.

The tranquil water of the Tarawera River belies the volatile origins of this part of the country.

Waioeka

Only 65 kilometres in length, the Waioeka descends from its source, close to the Motu in the rugged Huiarau Range in Te Urewera National Park, and widens out on the coastal plain, flowing to the sea via the estuary at Opotiki, an outlet it shares with the Otara River.

Its name is thought to be a corruption of Waioweka, meaning the water of weka. While that was true once, the birds disappeared from the region in the 1920s and only now are a few making a comeback.

Ancient pa sites near the river have been uncovered far inland and, from investigating the middens associated with them, it would appear that Maori made seasonal journeys into the interior seeking shearwaters, a surprising discovery when we think of shearwaters being birds of the coast and sea.

The Waioeka is a popular recreational river. As well as brown and rainbow trout, it is also home to native lamprey, koara (a variety of whitebait), eels and whitebait, and it was in the Waioeka in 1904, that one of the last grayling was taken. A native fish once common in many New Zealand rivers, it is now extinct. Wilderness fishing is possible near the headwaters of the Waioeka, but trout-fishing is accessible close to the main road near Opotiki.

S-bends near the mouth of the Motu River.

Wild and scenic: the Motu

Perhaps the most outstanding of these rivers is the Motu. Not major by New Zealand standards, it nevertheless has the reputation of being one of the country's most scenic rivers. A fishing website talks about its tranquil waters; a jet-boating pamphlet urges potential clients to ride the mighty Motu; a canoeist describes it as wild and scenic. No one would deny that this is a river of contrasts: a waterway of bouldery rapids, deep pools, and calm 'S' bends winding slowly between shingle banks.

It rises in the southwestern side of the Raukumara Range and flows into the ocean northeast of Opotiki. The upper reaches of the river run through lush farmland, the result of forest clearing in the early years of the 20th century, but a few kilometres downstream from the township of Motu the river plunges over the Motu Falls, crossed by a frail swing bridge high above the water. Kettle holes worn into the rocks beneath the bridge and debris caught in its railings show that the river has a wild side. For most of its 165-kilometre journey to the sea it rushes through steep-sided gorges in country clothed in dense native forest. In its last stage, the river opens out into long lazy loops. The highway south, which clings to the coast before it rounds East Cape, takes a winding detour up the Motu estuary before crossing the bridge and returning seawards, high above the meandering river. When we pass that way in early autumn, the gravelly riverbed is splashed with bright, silvery gold from masses of toetoe in flower, and great heaps of driftwood litter the beaches. We stop at a headland. In the distance fishermen try their luck where the river meets the sea.

Motu usually signifies 'island' in Maori, but it also means 'isolated' or 'severed', an appropriate name for a river that cuts through the last significant area of native bush in the North Island not traversed by tracks. Maori in the past, though, followed trails along the river's banks and cultivated fern gardens for seasonal food on terraces high above the water. The vegetation has long since re-established itself, but on the steep unstable country erosion is a continuing problem, caused by wild goats, pigs and possums ravaging the bush. Canoeists who have known the river for 30 years bemoan the shallowing of the riverbed over that period. In the upper reaches, future harvesting of plantation forests will add to the erosion and the resulting sediment load in the riverbed.

In 1983, the river was the first in the country to be put under the protection of a National Water Conservation Order, recognising its outstanding recreational values but also due in part to a population of about 80 native blue duck or whio that inhabit the Motu and its tributaries. Now an endangered species, they choose a habitat of unimpeded, fast-flowing water. The river is also home to the endemic Hochstetter's frog, brown trout and long-finned eel, and traditional eeling is carried out on the middle and lower reaches.

A popular canoeing venue, the Motu ranges from grade three to five depending on the flow, and reaches grade six when in flood – and it is prone to flash flooding that can happen in any month of the year. From the put-in site, not far from the township of Motu, it takes between three and five days to reach the Bay of Plenty, 90 kilometres away. On the way, tight, steep rapids become impassable during flooding.

Early canoe trips on the Motu were often more of an adventure than they are today. Two separate groups in 1919 and 1935 each took 10 days to make the voyage; both ran out of food and were met by rescue parties. It is also a favoured river for white-water rafting and jet-boating, and hunters use it to gain access to remote shooting blocks.

Those who prefer hiking over water sports can explore the Whinray Scenic Reserve, on the Motu Falls Road near the township of Motu. Set aside as a bush reserve at the beginning of the 20th century, the forest that remains there today pre-dates European settlement. In all it covers 430 hectares of bush running down to a spectacular set of waterfalls on the Motu. Fantails, kereru, tui, kaka and kiwi live there, and the reserve is home to a conservation project run by DOC that aims to protect North Island weka. These cheeky birds were once numerous throughout the North Island, but now the populations in the Motu and nearby Whitikau and Toatoa valleys are some of the remaining few left in the wild.

Multi-sports terrain

Each year, the area embraced by the Motu and the Waioeka is the scene of the Motu Challenge, a gruelling, multi-sport race that includes a run through the Whinray reserve, a cycling lap through the bush-clad Waioeka Gorge and a kayaking course down the Waioeka River. Another cycling lap on the Motu Road follows the route that Te Kooti was reputed to use, travelling between Poverty Bay and Opotiki during the uprisings of the early 1870s. In fact, a skirmish on the Waioeka in 1870 resulted in Te Kooti losing several of his followers, and his power diminished from that time.

At the beginning of the 20th century, attempts were made to open up the area for farming. Ballots for land were held in 1906, and later, soldiers returning from World War I were allocated land. Determined settlers struggled for years to make a living from the poor land but none of the farms was successful. By the 1970s all the failed sheep farms had been bought up by the government and they form the basis of the Waioeka Scenic Reserve.

Fishing at the Motu River mouth.

FISHING: Dry fly fishing for brown and rainbow trout on most rivers. Whitebait and kahawai fishery at Rangitaiki River mouth. Trophy fishing at Lake Aniwhenua.
WHITE-WATER RAFTING: Tarawera, Kaituna, Motu.
CANOEING: Tarawera, Kaituna, Motu, Rangitaiki.
SURF-CASTING: Rangitaiki River mouth.
OTHER ACTIVITES: Hunting, picnicking, swimming, jet-boating.

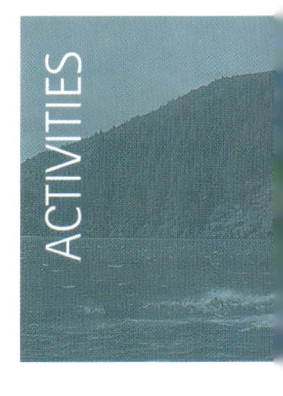

TAUPO – SLEEPING VOLCANO

The Tokaanu River flows around a small volcanic cone near the Tokaanu Power Station on its journey to Lake Taupo.

On a clear winter's morning, Lake Taupo sparkles in the sun. Across its vast expanse, the mountains in the distance are visible, tempting skiers. In summer the lake is abuzz with boats, and tourists drift around the town that clings to its shores. At any time of the year, golfers come to Taupo to enjoy world-class golf, and anglers arrive, keen to try their luck in the waters of the lake. Taupo's rainbow trout, introduced in the 1890s, draw visitors from all over the globe.

I first visited the lake when I was a child in the 1950s, when the town of Taupo consisted of several shops, a scattering of holiday homes and a few spartan fishing lodges. A village that had developed haltingly as a way-station for travellers between Auckland and Wellington, it boasted only a few accommodation houses. The lake shore lacked the manicured parks of today and was largely unencumbered by buildings. I remember my delight in finding a warm spring bubbling up on the beach where we went swimming, and lumps of pumice strewn over the gravel – unheard of wonders to a South Island child, accustomed to ice-cold lakes and pumice bought at the pharmacy. I also remember big breakers rolling in to shore, whipped up by a sudden wind, which made the lake look more like the sea.

Given the lake's origins, neither the hot springs nor the pumice should be a surprise. Lake Taupo is only about 27,000 years old, a mere infant in geological terms. It lies in an enormous caldera, formed by a massive volcanic eruption that spewed out poisonous ash, rocks, lava and pumice in enormous quantities and changed the face of the North Island. Entire forests were destroyed and, in some places, debris was deposited tens of metres deep. Several hundred square kilometres of land collapsed to form the present lakebed, the largest freshwater lake in Australasia.

As always, Maori have a much more personalised story to explain the origins of the lake and to give validity to their tribal claims. According to their legends, it was the tohunga from the *Arawa* canoe, Ngatoroirangi, progenitor of Ngati Tuwharetoa, who created the lake. After making landfall on the east coast, he led a reconnaissance party inland and on reaching the summit of Tauhara, they were dismayed to see an extensive barren basin spread below them. Ngatoroirangi seized a large totara tree and hurled it into the desert, planning to seed a new forest. But a fierce wind carried the tree off course, it struck the edge of the basin and landed upside down, piercing the crust of the land. Water gushed up, filling the

A LAND OF WATER

entire depression, and thus the great lake was formed.

Numerous eruptions have occurred since its creation – the most significant happening in AD 186 when the lake took on its present form. Estimates suggest the eruption column then was 50 kilometres high, twice as high as that of Mount St Helens in Washington State, USA in 1980. So great was the amount of ash emitted, that skies over Rome and China turned dull red. The most recent effect of volcanic activity was in the mid-1990s. When Mount Ruapehu erupted in 1995 and 1996, an estimated 2.3 million tonnes of ash was dumped into the lake, sinking to the bottom and creating a silted layer about four millimetres thick over a large part of the lakebed. The ash acted like a flocculent, used to clear swimming pools, and gathered sediment particles and algae together, taking them to the bottom. The water quality seemed to improve, but this is likely to be of limited duration because the trapped nutrients will gradually escape as the ash layer deteriorates and breaks up.

There is little evidence of early Maori settlement this far inland, although Ngati Tuwharetoa has been the main iwi in the area for several hundred years and temporary encampments have existed around the shores of Lake Taupo and Lake Rotoaira for generations. The island of Motutaiko, towards the southern end of Lake Taupo, was a Ngati Tuwharetoa stronghold in the 17th and 18th centuries, a place to retreat to in times of trouble. Several caves on the northern side of the island contain ancient burial sites and are still considered sacred and tapu today.

Not until about the 1840s, when major pa were established at the southern end of the lake, was there any significant Maori population. Missionaries arrived about the same time, but the converts were few and the mission was short-lived. European settlers did not arrive in any numbers until 1869, when Taupo became a garrison town in response to the skirmishes involving Te Kooti. The government bought the site on which the town now stands, but later attempts to farm the land failed miserably because of a cobalt deficiency in the soil. It took masses of money and labour to first clear it and then apply fertiliser before farming became viable. In addition, winter conditions were harsh, the roads in the general area were appalling and, difficult to believe now, the township did not have its own electricity supply until 50 years ago.

It was only in the 1950s that the Taupo region began to flourish. Farming took off on the rehabilitated pasture, forestry burgeoned and the construction of hydroelectric and geothermal power stations nearby brought a new industry to the area. Nowadays, not only is Lake Taupo the reservoir that feeds the hydroelectric system along the Waikato River, but it is also an important tourist centre, catering for large numbers of national and international visitors, with an extensive shopping precinct and endless accommodation options. Some people come to relax in luxury; others to indulge in a range of adventure pursuits or endurance events such as the New Zealand Ironman contest.

TOP: The rock carving by Greg Whakataka Brightwell on the shores of Lake Taupo can only be seen from the water.

ABOVE: Motutaiko Island, in Lake Taupo, has long been a sacred site for Maori.

BORN OF VOLCANOES   49

ABOVE: Lake Taupo, a caldera of a huge ancient volcano, looking towards its sisters, Mount Ruapehu on the left and Mount Ngauruhoe on the right.

RIGHT: Renowned for its trout-fishing, the Tongariro River flows into Lake Taupo.

A LAND OF WATER

A hint of autumn:
Moutere Point, on the
eastern shores of Lake
Taupo.

BORN OF VOLCANOES

Isolated villages have grown up around the lake's periphery, many dating from pre-European times, some the site of natural geothermal activity. Waihi is one. An historic Maori village, it sits at the lake edge near Tokaanu. Fumaroles are present on the cliffs above Tokaanu, while at Waihi there is an area of steaming ground and hot springs that, at both locations, have long been used for domestic purposes and recreational bathing.

Tokaanu was the largest Maori community at the southern end of the lake when Europeans arrived in the area. With its mud pools, hot springs and a geyser, and its proximity to the famed Tongariro fishing pools, it is now a popular holiday location as well as the site of one of the power stations on the Waikato River.

In common with many other waterways in New Zealand, concerns have arisen about the quality of the lake water and how it is affected by land usage around it. Environment Waikato issued a strategy setting out guidelines for protection of the lake with a focus on reducing and capping the amount of nitrogen leaching into the water from both rural and urban sources.

RUAPEHU – CRUCIBLE OF DISASTER

Breathtakingly beautiful and potentially deadly, Crater Lake at the top of Mount Ruapehu is New Zealand's largest cone volcano. Its history is one of continuing activity. Over the last 50 years alone there have been numerous eruptions – several occurring in the 1970s emitting ash and causing lahars that damaged the nearby skifields.

The last eruptions in 1995 and 1996 emptied Crater Lake, but the water level within the crater rose steadily again (between April 2001 and March 2004 it rose by 23 metres) with a layer of loose debris called tephra (ash, sandy particles and larger material) forming an unstable dam in the crater rim. On 18 March 2007, the water level became too high and this

ABOVE: The Tangiwai Memorial at the site of New Zealand's worst train disaster.

RIGHT: The raw crater of the cinder cone of Mount Ngauruhoe contrasts with the brilliant blue of the Tama Lakes in the distance.

barrier eventually collapsed. As a result, a moderate-sized lahar, a flowing mixture of mud and volcanic debris, passed down the Whangaehu River, luckily with minimal damage to infrastructure in its path, or disruption to the travelling public.

A similar lahar was, however, responsible for the worst train disaster in New Zealand's history. On Christmas Eve 1953 a huge wall of water, mud and rocks poured down the Whangaehu River, sweeping away a concrete support of the rail bridge at Tangiwai and collapsing the bridge minutes before the arrival of the north–south train. Unable to stop in time, the train plunged into the river, killing 151 passengers.

Given that this is a national park and there are no areas of dense population close by, a decision was taken not to intervene in future collapses of the Crater Lake by engineering means. As the 2007 experience shows, inevitably more lahars will happen, but the steps already taken were instrumental in preventing a disaster similar to Tangiwai. Stronger bridges have been built and warning systems installed, and response plans developed to safeguard people in the potential path of a lahar. Vulcanologists will also continue to watch the lake levels, water temperatures and any other signals that might indicate violent activity.

Photographs of Crater Lake taken in the 1950s used to feature bikini-clad swimmers lounging at the edge of the lake, their skis poking out of the snow beside them. Alas, swimming in Crater Lake between ski runs is no longer possible as the water has become too acidic. But it is possible to climb to the lake, from where there are spectacular views of a large area of the North Island.

The stream at Tangiwai and the bridge behind, which was the site of the Christmas Eve disaster in 1953, when a lahar from Ruapehu's Crater Lake swept it away just before the arrival of the main train heading north, resulting in the death of 151 people.

BORN OF VOLCANOES

The Crater Lake of Mount Ruapehu, an active volcano, with Mount Taranaki on the horizon.

OPPOSITE: The waters of a typical stream descending from Mount Ruapehu to Lake Taupo.

54 A LAND OF WATER

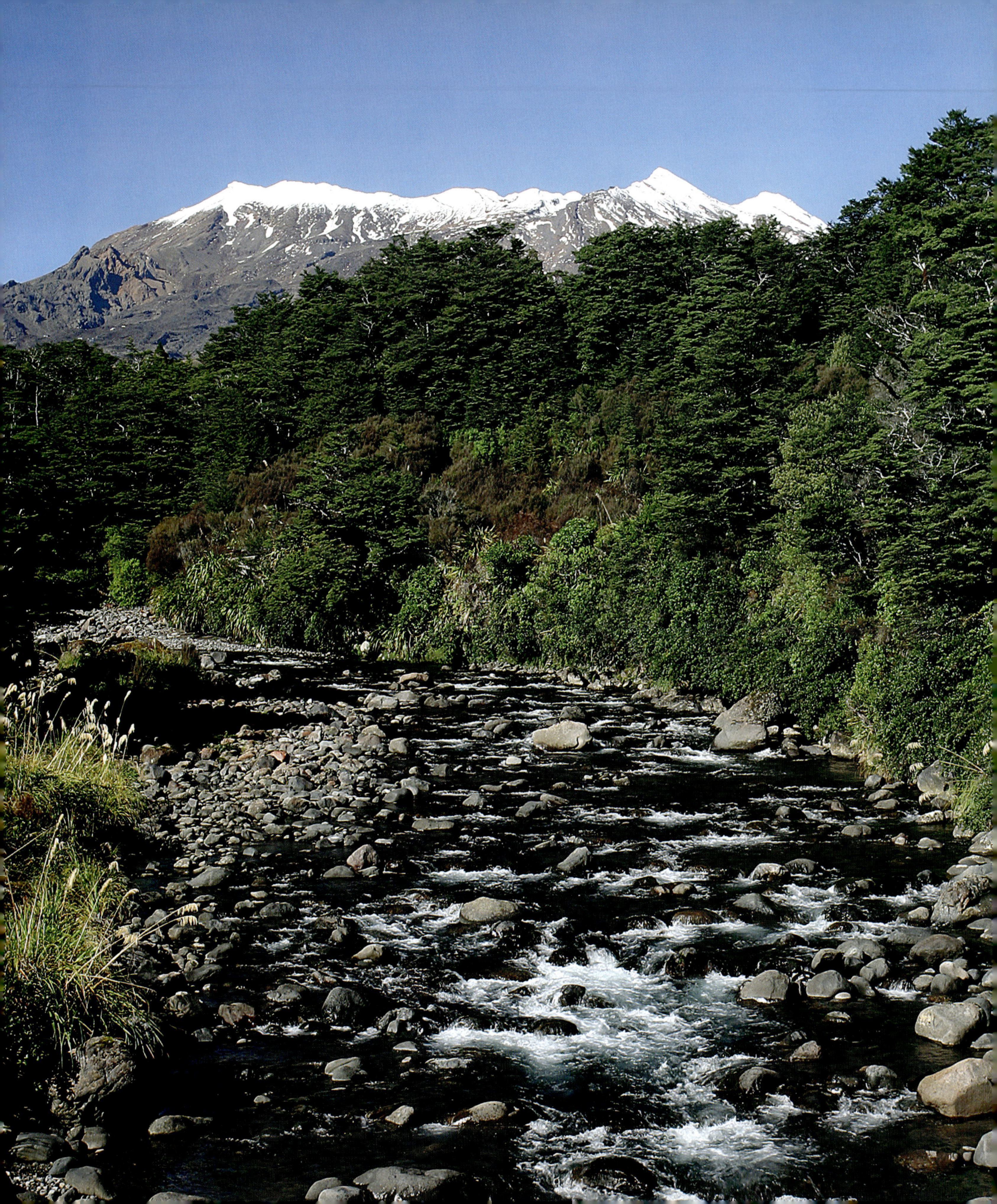

FROM CAPE TO CAPE

EAST COAST WATERWAYS

The Waiapu River passes through the town of Ruatoria and discharges into the sea just south of East Cape.

The land on the East Coast of the North Island is fragile – so fragile that hillsides erode at an appalling rate in the heavy storms that occur frequently in the area. Huge quantities of gravel and sediment wash down into the rivers, filling up their beds, threatening the bridges and washing out to sea every year. But drive south on State Highway 35 from Te Araroa near East Cape through the rural farmland north of Gisborne on a good day and you have the feeling of travelling in a benign countryside.

In early autumn, exotic trees are turning golden. The air is warm. Gravelled river braids glitter in the sun. Yet this part of the country has changed, often violently, and continues to change year by year.

Even before humans came on the scene, the east coast was prone to erosion because of its geologic make-up. Composed of rock types that tend to make a slurry when water penetrates deep within them, they slough off in heavy rain, creating landslides that, over the millennia, have been aided by earthquakes. The region accounts for a third of the soil washed into the sea from New Zealand rivers each year. It attracts international scientists, who come here to analyse the sediment in the ocean, in an effort to chart important geological and climatic happenings over past centuries.

When humans arrived and started burning trees, and later chopping them down, to create grazing land, the breaking down of the hill country accelerated beyond anyone's imagination. Fast-flowing in their headwaters, many of the rivers in the Gisborne area are slow-moving once they leave the hills, flowing through wide valleys. Rivers meander in this kind of territory and, with the denudation of the hills, wide shingle beaches build up in the curves of the rivers. After rain, they become laden with silt, and gradually fill with rubble. As the riverbeds become shallower, so the risk of flooding increases.

In the 1960s, in a concerted effort to ameliorate the situation, the Forest Service acquired land and began planting trees on a large scale. At the same time, the Gisborne Catchment Board adopted a policy of advocating farm planting. Pine trees were recommended because of their superb efficiency in soaking up water. Today exotic forest, mainly pines, occupies 26 percent of the catchment, while indigenous forest still covers 21 percent, mainly in the steep mountain ranges, where high rates of natural erosion occur. Sometimes nature, however, seems determined to undermine the gains that humans have belatedly made. When Cyclone Bola hit the coast in March 1988, more than 900 millimetres of rain fell in five days, sluicing soil, gravel and rocks off the hills and wiping out houses, bridges and roads. The damage bill was close to $200 million.

Erosion control is an ongoing concern. The East Coast Forestry Project, funded by government, is paying farmers to plant their land in trees, and although much progress has been made, there is still much land that remains exposed to erosion.

A similar story has unfolded in Hawke's Bay where three rivers – the Ngaruroro, the Tutaekuri and the Tukituki – have significantly shaped the boundaries of Napier and Hastings. From the ranges to the coast they have laid down rich soil, which has played a large part in the prosperity of the area and helped to create an environment in which today so many of New Zealand's grapes grow. But these rivers too have carried a heavy load of debris down from the hills. Used by Maori for centuries as waterways for voyaging canoes, they were still navigable in the 19th century. Today that is no longer true and, here also, erosion is an ongoing problem. The riverside areas are managed primarily for flood control purposes but they have also been enhanced to provide areas for picnicking and fishing as well as trails for walking, biking and horse-riding.

Maraehara and Waiapu rivers

Meandering and slow in its lower reaches, the Maraehara River starts life in a forested area in the hills to the west of East Cape and joins the Waiapu at its mouth where it flows into the Pacific Ocean between East Cape and Whakariki Point north of Ruatoria. The Waiapu itself follows a course through a wide valley on the coastal plain, but tributaries such as the Tapuaeroa and the Mangaoparo start life in the hilly areas to the west.

European settlers came to this area about the 1880s, and pastoral farming started in the Waiapu catchment soon afterwards, initiating a phase of greatly increased erosion and sediment transfer. In one of the world's worst examples of erosion, the Waiapu River delivers approximately 20,500 tonnes of soil and gravel into the sea each year for every square kilometre of its catchment, this in spite of efforts that have been made to improve the situation. So great is the aggradation of the river that several bridges across it need raising.

Turanganui River

Gisborne, straddled by rivers and sometimes called the 'City of Bridges', is situated on the shores of Poverty Bay where the Taruheru and Waimata rivers join to form the Turanganui River. Long ago, the Turanganui River flowed out to sea to the northwest and there was a natural channel between the river and a nearby reef that Maori used for centuries, to provide canoe access to the river. Now it is known as the shortest named river in New Zealand, being only 1200 metres in length. At its mouth is a statue of Nicholas Young, reputed to have been the first crewman on board the *Endeavour* to sight land and whose name was given to the white cliffs across the bay. Interestingly, Kaiti Beach, near the city, is traditionally recognised as the place where the *Horouta*, one of the ancestral Maori waka, landed and it is also where Captain Cook first set foot on New Zealand soil in 1769.

A gracious Riverside Walkway, planted with some fine old trees, winds its way around the banks of the Turanganui, adding panache to the city and serving as a fitting site for two statues that commemorate important Gisborne citizens. The Margaret Sievwright Memorial remembers an early suffrage worker. Wi Pere Memorial on Reads Quay honours a Maori leader who was a Member of Parliament from 1884–1912.

Autumn on the Maraehara River, East Cape.

ABOVE: The Waipaoa River drains the country behind Gisborne and flows out to sea at Young Nick's Head.

RIGHT: The lighthouse relocated to the main street in Wairoa from nearby Portland Island.

Waipaoa River

According to Maori legend, Paoa, an important chief in times gone by, lived in the remote interior of the East Coast before the river existed. He built a canoe and then needed a river where he could launch it. So he set about creating the Waipaoa.

It begins in the heavily forested Raukumara Range, a rugged area rising to more than 1000 metres, and is joined by several tributaries before it reaches the lowlands north of Te Karaka. As it wanders seawards, it picks up yet more smaller rivers, all draining into the main river valley from the hills to the east and west. After meandering over a wide estuary area the river empties into the sea between Gisborne and Young Nick's Head.

A large proportion of the total catchment area is soft, clay rock. With the intensive downpours that plague the region, this material forms a thick paste with water and flows off down the hillsides causing a very high sediment content in the Waipaoa. Consequently, the riverbed is continually building up, causing enormous problems in river control. River straightening works have been carried out to try to minimise the effects of flooding, and stopbanks have been built to protect the city.

Lake Waikaremoana

Neither glacial nor volcanic in its origins, Lake Waikaremoana was formed approximately 2200 years ago, when a colossal landslide blocked the course of the Waikaretaheke River and created a massive dam. It is estimated that the landslide blocked the river with debris over 300 metres thick, spilling up and down the valley for three kilometres. The result was Lake Waikaremoana, 'Sea of Rippling Waters', one of the larger lakes in the North Island. It is situated on the southeastern boundary of Te Urewera National Park and sits among remote, bush-clad mountains between Rotorua and Wairoa. Tuhoe, known as 'The Children of the Mist', is the Maori tribe of the area and some groups still retain a more or less traditional way of life in the isolated valleys.

A second landslide followed the first and settled in an almost intact block on top of the earlier debris. As this block of layered sandstone moved and fractured, the Onepoto Caves were formed. A series of rock overhangs and caverns, they include tunnels as long as 20 metres, some with multiple entrances. Maori legends tell of the might of the warrior Tuwai, who was ambushed by five attackers as he slept in one of the caves but who nevertheless managed to emerge victorious.

Magnificent scenery and the opportunities for good swimming and fishing attract large numbers of visitors each year, many of whom hike the Waikaremoana Great Walk, a 46-kilometre, three-to-four-day tramping track that follows the shore of the lake for most of its length.

Nearby is Lake Waikareti, also formed by landslide action. One of the most beautiful forest walks in the Urewera Range climbs through beech forest to reach this idyllic lake with its crystal clear waters. Boasting exceptionally high water quality, it is the largest of

A LAND OF WATER

Lake Waikaremoana from the air.

the few lakes in the North Island that remains free of introduced aquatic weeds, making it an important control for research on similar lakes. The surrounding forest is rich in native birds such as tui, bellbirds, kereru (New Zealand pigeons), parakeets, kaka, riflemen, robins and fantails.

The hydroelectric generation potential of the area was recognised towards the end of World War I, and a local scheme for Wairoa was completed in 1923. Tuai Power Station was opened in 1929, Piripaua in 1943 and Kaitawa in 1948. Waikaremoana has provided water storage for the hydroelectric power stations on the Waikaretaheke River since 1929. Lowered in 1946 to try and decrease natural leakage and to increase the water flow through the Kaitawa Power Station, the lake now operates at an average of five metres below its natural level and the Department of Conservation is involved in monitoring the resource consent to ensure that the hydroelectric use of the lake minimises excessive lake level variations.

Wairoa River

Much of the catchment of the Wairoa River is densely forested and is a little-known area, difficult to penetrate. As it winds out to the coast south of Gisborne through gentler land, this, the biggest river in Hawke's Bay, is the culmination of several tributaries, including the Waikaretaheke, whose dammed headwaters formed Lake Waikaremoana; the Waiau, which rises in the Kaiangaroa Forest; and the Ruakituri and the Hangaroa, both of which rise in the north, not far west of Gisborne. Studded with willows and poplars, the banks of the river turn tawny gold in autumn.

Maori are believed to have first arrived in the district around 1300, one of their early settlements being on the banks of the Wairoa River, not far from the present location of the ornately carved Takitimu Marae, named for the great waka, *Takitimu*. Known as Te Wairoa, meaning the Long River, it was inhabited by Ngati Kahungunu.

The mouth of the river was significant enough for Captain James Cook to note it on his chart during his first visit to New Zealand in 1769.

European visitors did not arrive in Wairoa until the 1820s, when flax traders set up in business. In the next decade a trading and whaling station was established there and a few years later missionaries arrived, including William Colenso who travelled extensively throughout the eastern North Island. During the 1850s the town developed as a river port, growing up on the southern bank of the river, about two kilometres from the mouth. Flax, fruit and timber,

The Upper Mohaka River.

produced in the surrounding area, were shipped to Napier. Later, wool, meat and dairy products were also exported.

As with other rivers on the East Coast, the Wairoa has been prone to serious flooding. In 1948, the Wairoa traffic bridge was submerged and parts of the town were a metre deep in water. Cyclone Bola swept the bridge away and the present one was completed in 1990. Today, the town's most distinctive landmark is the relocated Portland Island lighthouse. It stands on the banks of the river, a dramatic if unexpected landmark in the middle of town. Nowadays a centre for many water-based leisure activities, Wairoa is also a gateway to Te Urewera National Park and Lake Waikaremoana.

Te Reinga Falls

Te Reinga on State Highway 36 is situated near the confluence of the Ruakituri and Hangaroa rivers, which together form the Wairoa. Nearby are the magnificent Te Reinga Falls – silky, smooth sheets of water that break up and tumble over rock shelves into inky deep pools. As early as 1904 it was suggested that their water could be harnessed to produce hydroelectric power. Investigations continued intermittently over the years, but fortunately any schemes were subsequently abandoned and the falls have been allowed to retain their natural grandeur.

Shell fossils, surprisingly, are easy to find, encrusted in flat slabs of rock near the water, suggesting that at one time the area formed part of the sea floor. In 1973, bones of an unknown species of penguin, found embedded in calcareous sandstone, reinforced this idea. Fragments of dolphins, whales, penguins and seals have also been found in the area.

This is the ancestral domain of Ngati Hinehika. They ascribe the formation of the falls to two taniwha, a brother and sister, who lived high in the forested hills. They decided to race each other to the sea. Hinekorako carved out the Hangaroa River; Ruamano formed the Ruakituri River. Hinekorako was the first to arrive at Te Reinga. Relaxing after her rigorous journey, she sat on a huge rock, combing out her luxurious, long hair. Ruamano saw her as he approached and carved a new path over the falls, stealing a march on her in the race to the ocean at Wairoa. Hinekorako became aware of the water receding behind her, realised she had been beaten and decided to remain where she was.

Mohaka River

Some claim the Mohaka is the last wild and scenic river in the North Island. While others would no

A LAND OF WATER

doubt dispute this, no one would argue that the river and some of its tributaries possess outstanding scenic characteristics that make it an important area for water-based recreation, a fact recognised in 2004 when the Mohaka was granted the protection of a Water Conservation Order. Among the special features cited were an outstanding trout fishery, the beauty of the Mokonui and Te Hoe gorges and the river's importance for such sports as rafting and canoeing. Boulders the size of houses and steep cliffs rising 150 metres above your head make the Mohaka a wild and scenic river for those who like adrenalin-pumping water rides.

The river that travellers see from State Highway 2, between Wairoa and Napier, is a very different one from the upper reaches accessed by adventure seekers in the Kaweka Forest, close to its source. As it approaches the coast, cultivated paddocks spread atop perpendicular cliffs, which the river has cut through over time. At their base, the river is a moss-brown strip, shallow and pebbled, with no hint of the hectic raft ride it offers further inland.

The highest railway viaduct in Australasia was built on the Mohaka's lowest reaches in 1933. It remains as a testament to the skill of the engineers involved in its construction – and as a challenge to photographers who feel the need to get up close and personal with it.

Known for its fishing potential, the Mohaka became headline news in the mid-1980s. An amateur palaeontologist called Joan Wiffen, from Haumoana in Hawke's Bay, discovered a tiny bone in a tributary of the Mohaka, which palaeontologists decided came from a carnivorous dinosaur called a theropod. It probably stood about two metres high and weighed slightly less than half a tonne. Since then, Joan Wiffen and her colleagues have come up with several other dinosaur species not previously found in New Zealand.

Lake Tutira

North of Napier, Lake Tutira, with its tiny neighbour Lake Waikapiro, are both idyllic spots for picnicking and camping. More than 60 years ago the area was declared a bird sanctuary at the instigation of William Herbert Guthrie-Smith, an ardent conservationist and ornithologist who farmed neighbouring Tutira Station until 1940. Now the station is owned by the Guthrie-Smith Trust, and walkways pass across its land.

We park our camper van one evening on the shores of Lake Tutira, close by a couple of swamp cypresses, with their distinctive knobby roots. A flotilla of black swans keeps watch. Their long necks form graceful question marks across the water, when they're not busy diving. One pair though has mating on their minds. Larger than the others, they have commandeered a corner for their courting ritual. White wing feathers peep out beneath their black tutus and their stately minuet draws raucous squawks from ducks in the cheap seats.

ABOVE: The viaduct over the Mohaka River near Raupanga.

LEFT: The view beneath the Mohaka Bridge.

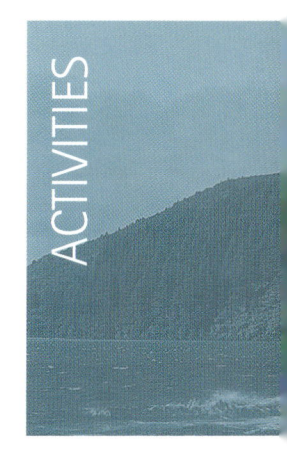

RAFTING: The Mohaka/Waipunga River run is rated as one of the top ten trips in the North Island. Commercial trips on the upper Mohaka start from the boundary of Kaimanawa Forest Park and Poronui Station, or from the confluence of the Mohaka and Taharua rivers.

FISHING: The Mohaka, Makino and Ngaruroro rivers, where rainbow and brown trout are known for their size and fighting ability.

CANOEING: The Wairoa is a popular paddle from Te Reinga Falls to Marumaru Hotel, over numerous short ledge rapids with dense native bush lining the riverbank.

ACTIVITIES

Lake Tutira.

of how the land is affected by changes in climate. Facilities at Lake Tutira include a bird sanctuary, a wildlife refuge, walking tracks, toilet and picnic facilities, campsites, fresh water and barbecues.

Rivers that embrace Napier

The Tutaekuri, fed by rainwater from the Kaweka Ranges, used to flow out through the estuary behind Napier Hill before the 1931 earthquake. Now it has a common mouth with the Ngaruroro and Clive Rivers – an area that provides good feeding and nesting areas for migratory birds. The river's name, which means offal of the dog, originates from a long-ago incident. When a group of Ngati Kahungunu from Wairoa were on their way home, they arrived at Hikawera's home between Omahu and Waiohiki. He had cooked 70 dogs to feed the travellers, and the offal was thrown into the river.

In the morning we awake to see a snooty goose high-stepping past the van, checking out her territory. A solitary swan punctuates the clouds of vapour steaming off the water. Frogs croak. A heron picks its way disdainfully through the reeds.

A couple of ancient, converted buses rumble in, pulling up with a flourish side by side. 'Stone Age' reads the sign scrawled across one cab. A door springs open and there is a sudden commotion. Out tumbles a small boy followed by a couple of boisterous dogs and more children, six in all. The goose takes hurriedly to the water, her dignity in tatters. Father appears and tinkers with one bus; mother emerges from the other holding a baby. Home on wheels or holiday adventure? Whichever, I admire their courage.

Today's refugees from the city are not the only people to appreciate the peace of Lake Tutira. Centuries ago, Maori established settlements around its edge and its history is ancient. Formed originally as the result of a landslide, the lake has recently yielded up secrets of life on earth long before people roamed its shores. In 2003, the National Institute of Water and Atmospheric Research (NIWA) took a 26-metre core sample of sediment from the lake bed to study weather patterns over the last 7000 years and so gain a clearer picture

Also nurtured in the Kaweka Ranges is the Ngaruroro River which supplies water to the many farmers and horticulturists on the Heretaunga Plains. It too has a picturesque name. Ngaru means ripple, and upokororo was the name for a freshwater fish which is now extinct. Early in Maori history, Mahu was on his way home to Mahia when his dog startled a shoal of upokororo, causing the water to ripple.

One of the two major rivers flowing across the Ruataniwha Plains in central Hawke's Bay is the Tukituki River. It flows north and east from the Ruahine Ranges, combines with the Waipawa River downstream from the towns of Waipukurau and Waipawa, and exits into the ocean between Clive and Haumoana. During the 1880s, in the early days of farming, barges used to travel down the river, carrying wool from Waipawa to deliver to lighters off the coast at Haumoana. Its name comes from a fishing technique used by Maori. Tuki upokororo means to slap the water to drive small fish into the net.

OPPOSITE TOP: Lake Onoke with Lake Wairarapa in the background.

OPPOSITE BOTTOM: A cable-stayed footbridge over the Hutt River near Upper Hutt.

WATERWAYS OF WAIRARAPA AND WELLINGTON

Large amounts of water regularly sweep down onto the Wairarapa Plains from the steep, aggressive rivers that pour out of the surrounding bush-covered mountains to the west and the hilly country to the east. But once they have arrived on the flat lands, the waters slow down and spread out in a network of streams, lakes and lagoons that provide irrigation and stock water during the often dry Wairarapa summers.

Most of the rivers, such as the Tauherenikau, find their way into Lake Wairarapa, a shallow wetland area rarely more than two and a half metres deep, situated just south of the town of Featherston.

The course of the wandering Ruamahanga, however, which rises in the Tararua Range and finds its way south in loops and bends, has been altered by flood alleviation works. It drains into Lake Onoke, which borders the ocean at Palliser Bay. While farmers have benefited from the flood protection, many small wetlands have disappeared, along with their plants and animals.

Waterfowl hunting has been a traditional activity in the wetlands since the first birds were introduced to Lake Wairarapa in the 1880s, and hunters vehemently opposed plans to drain the eastern side of Lake Wairarapa for farming in the 1960s. Although the wetlands have been reduced by almost half since then, the area is still important for the wildlife it attracts. Some migratory birds travel more than 10,000 kilometres to get to Lake Wairarapa and it is a favoured location for bird-watchers.

Duck-shooters, too, come from Wellington and Manawatu during the season in pursuit of paradise shelduck, grey duck, mallard, shoveller, pukeko, black swan and Canada geese.

The diversion of the Ruamahanga River was equally unpopular with Maori from South Wairarapa who, since time immemorial, have harvested the eels that migrate through the wetlands on their way to the ocean and their distant breeding grounds. The opening of the lake mouth in autumn, when the eels run, interferes with traditional fishing methods.

Sited as it is on the edge of Cook Strait, Wellington is more often associated with the sea than with any rivers. But when European settlers arrived in the area their first makeshift town was established beside the Hutt River at Petone. Severe flooding, however, soon persuaded the settlers to move to the Thorndon area. Today, the Hutt River sweeps through the cities of Upper and Lower Hutt and occasionally threatens them with severe flooding.

The Hutt River rises in the heavily forested southern Tararua Range and is fed by several tributaries – the Pakuratahi and the Mangaroa, which rise in the northeastern Rimutaka Ranges, and the Akatarawa and Whakatikei rivers, which flow into the Hutt from the hills to the east. Running alongside the Hutt River from Petone to Birchville in Upper Hutt is the 29-kilometre Hutt River Trail, an easy scenic walk and cycle path. It also allows access to the river for swimming, fishing and kayaking. The trail will eventually run as far as the Te Marua area of Kaitoke Regional Park, to link with the Rimutaka Rail Trail.

ACTIVITIES

Rivers of this area draw sports enthusiasts to enjoy the recreational opportunities they offer so close to the capital city.

RUAMAHANGA RIVER: Trout-fishing, kayaking, jet-boating, water-skiing.

HUTT RIVER: Swimming, kayaking, white-water rafting, jet-boating, trout-fishing.

LAKE ONOKE/LAKE FERRY: Surf-casting, traditional eel harvesting, whitebaiting.

LAKE WAIRARAPA: Wind-surfing, power-boating, water-skiing, yachting.

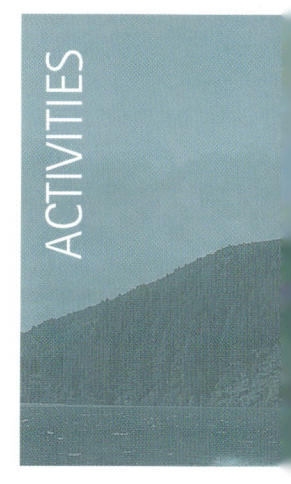

RIVERS OF THE WEST

TARANAKI

ABOVE: The Whanganui River Gorge, home to many Maori legends.

OPPOSITE: From mountain streams, the Waitara River eventually becomes a substantial and slow-moving river.

Taranaki is known for its stands of native bush and glorious rhododendrons, nourished by the region's temperate climate and regular rainfall. Less well known are the numerous streams and rivers of Taranaki, also a product of its regular rainfall. Many of them pour down the sides of Mount Taranaki like custard down the sides of an upturned plum pudding.

Tumbling and turning, they wind their way through dense rainforest and boulder-filled beds before continuing to the sea through green paddocks and endless humps of solidified lava. And when the roads that follow the two routes around the perimeter of the mountain were constructed all these streams had to be bridged. Imagine then how it was for Maori, making endless water crossings on their way north and south for centuries before the arrival of engineers.

These are not the long tortuous rivers so common in other parts of the country, although the Patea describes a convoluted scribble across the map after it leaves Stratford, while the Waitara, rising in the east, traces some deep loops as it makes its way to the North Taranaki Bight. Nor are there records of epic journeys on these waters, although the area is steeped in history. Pukerangiora Pa, high above a bend in the Waitara River a few kilometres upstream from the mouth, played a crucial part in the tribal wars of the 1820s and 1830s, and later, during the land wars that wracked the country in the 1860s, laying the seeds of dispute for the next 140 years.

In the South Taranaki Bight where the Patea River flows into the Tasman Sea, its tidal waters form an estuary. For about one hundred years it was a useful port for small coastal boats. Now a dam on its waters forms the rambling Lake Rotorangi, favoured by water-skiers and trout-anglers, and by eels in the river above. The associated power station feeds into the national grid, but not without making concessions on water flows on occasions, to allow migration of the mature eels downstream. A special pass was constructed alongside the dam to allow immature eels or elvers to pass up the river.

Egmont National Park, with a volcano in its midst, riddled with rivers and cloaked in lush native forest, is a place where visitors can enjoy a microcosm of almost everything that is typical of our country. There are numerous walking tracks that wind their way through the park, some for casual walkers, some for more experienced outdoors people. They offer the quiet beauty of deep river pools, waterfalls, wide views over a rural landscape and serious climbing in a rugged and sometimes dangerous environment.

There is something magical about waterfalls, and Dawson Falls on the Kapuni River are no exception. Only a short walk from the road near the Dawson Falls Visitor Centre in Egmont National Park, they are within easy reach of visitors. The track winds through gnarly old kamahi draped in moss, and as you pick your way down towards the river, the sound of rushing water gradually increases. It was the sound of the falls that drew Thomas Dawson to them in 1883 when he was exploring the mountainside; but he did not actually see them until several months later when he returned with two companions and they came across the waterfall that Maori must have known about for centuries.

Taranaki's rivers

Although flooding is common in the narrow valley floors of the hill country, Taranaki does not have any large flood plains. Flood control works are relatively minor throughout the region though stopbanks have been constructed to reduce the risk of flooding in the town of Waitara.

Ignored by the rest of New Zealand for a long time because of its geographical isolation, Taranaki is increasingly attracting people intent on water-based adventure and fishing in its more than 40 trout-fishing streams.

In its upper reaches the Waitara is popular for white-water rafting, but it arrives in town as a wide slow-moving river, popular for fishing, boating and whitebaiting, in season. The Waitara Bridge is known for the slightly mad venture of bridge swinging. Koi carp, recently found in the river, are a concern because of their potential threat to whitebait populations. At one time the river port handled the output of the

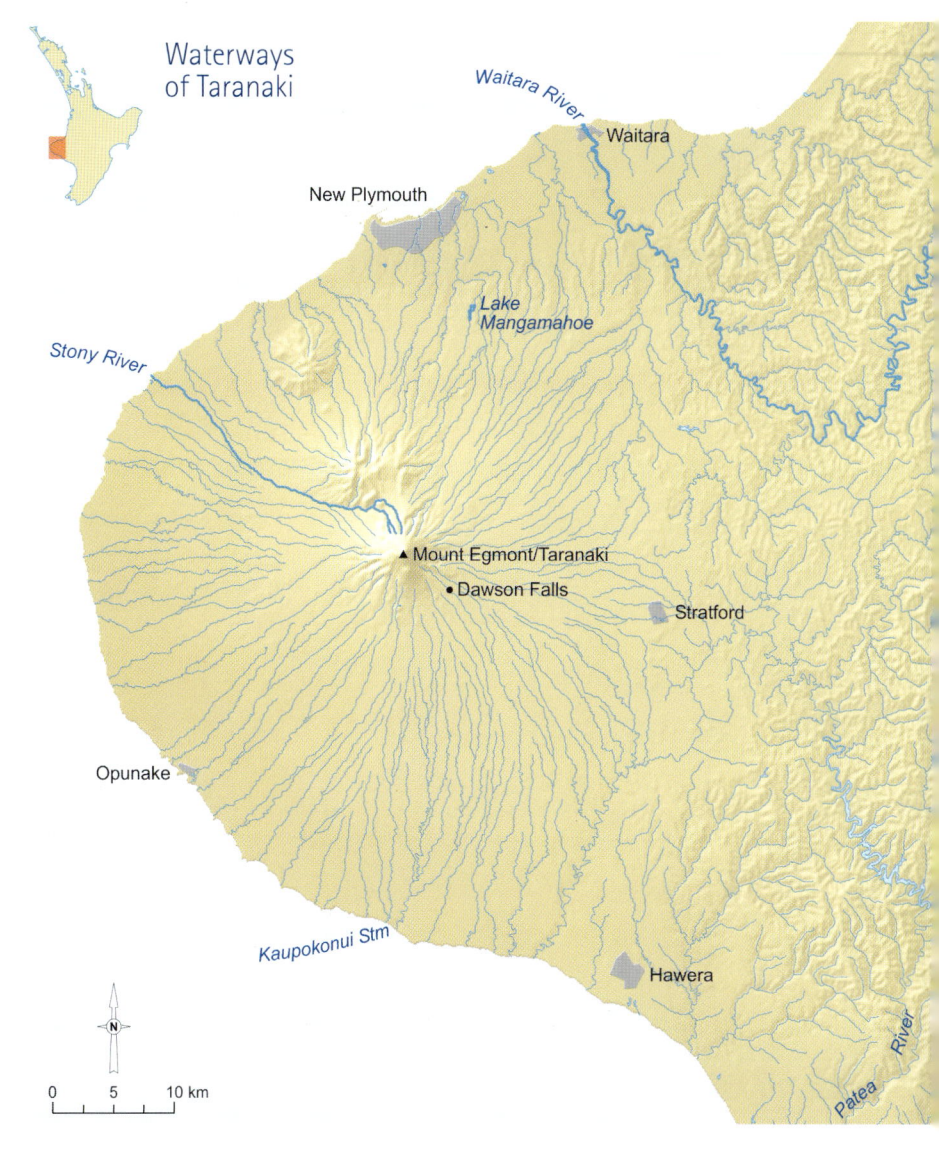

The Stony (or Hangatahua) River is one of the many that radiates from Mount Taranaki.

Complicated patterns of water and sand at the mouth of the Awakino River.

Waitara Freezing Works, but this meat now passes through the port of New Plymouth, and Waitara serves only small boats.

The Blue Rata Scenic Reserve is located on the Stony or Hangatahua River. It's a great place for a picnic, with beautiful clear swimming holes full of rainbow and brown trout. Starting from Old South Road in Okato, a scenic walkway meanders along the Kahihi Stream and the Stony River, through native bush and farmland. Only four kilometres long, it's a pleasant ramble and there are excellent views of Mount Taranaki.

Over the years many boats have plied the waters of the Mokau River. Maori used the river extensively, and later it served as a highway to the farms, coal mines, sawmills and lime kilns in the district. MV *Cygnet* was a stalwart of busy past days. Built in 1913, the vessel transported stock, farm and medical supplies, mail and household goods to people who lived and worked along the river.

On her return journey she would stop and collect cream from dairy farms and transport it to the dairy factory near the river mouth. With its closure in the 1950s, river traffic came to an end. Now, after years of neglect, the MV *Cygnet* has been restored to her former glory and takes tourists on trips up the river.

Recent plans to dam the Mokau for hydro-electricity have brought an angry reaction from canoeists. The river is recognised nationally for its wild beauty and white water, which paddlers have enjoyed for decades.

At the very northern limit of Taranaki, the Awakino River twists and turns its way through native bush in a series of rapids and deep pools that make it a favourite haunt of canoeists and trout-anglers. Walking tracks traverse its catchment area and State Highway 3, from the coast through to Te Kuiti, follows the river's winding route for several kilometres before it branches east towards Piopio. At its mouth, the river makes a long lazy loop before it flows out to sea, close to the mouth of the Mokau. This is a favourite river for whitebaiters and their typical huts sit at the edge of the river awaiting the rush of the fish.

ABOVE: Dawson Falls on the southern flank of Mount Taranaki, the rocks showing strata from various volcanic eruptions.

LEFT: Low tide on the Mokau River.

RIVERS OF THE WEST   67

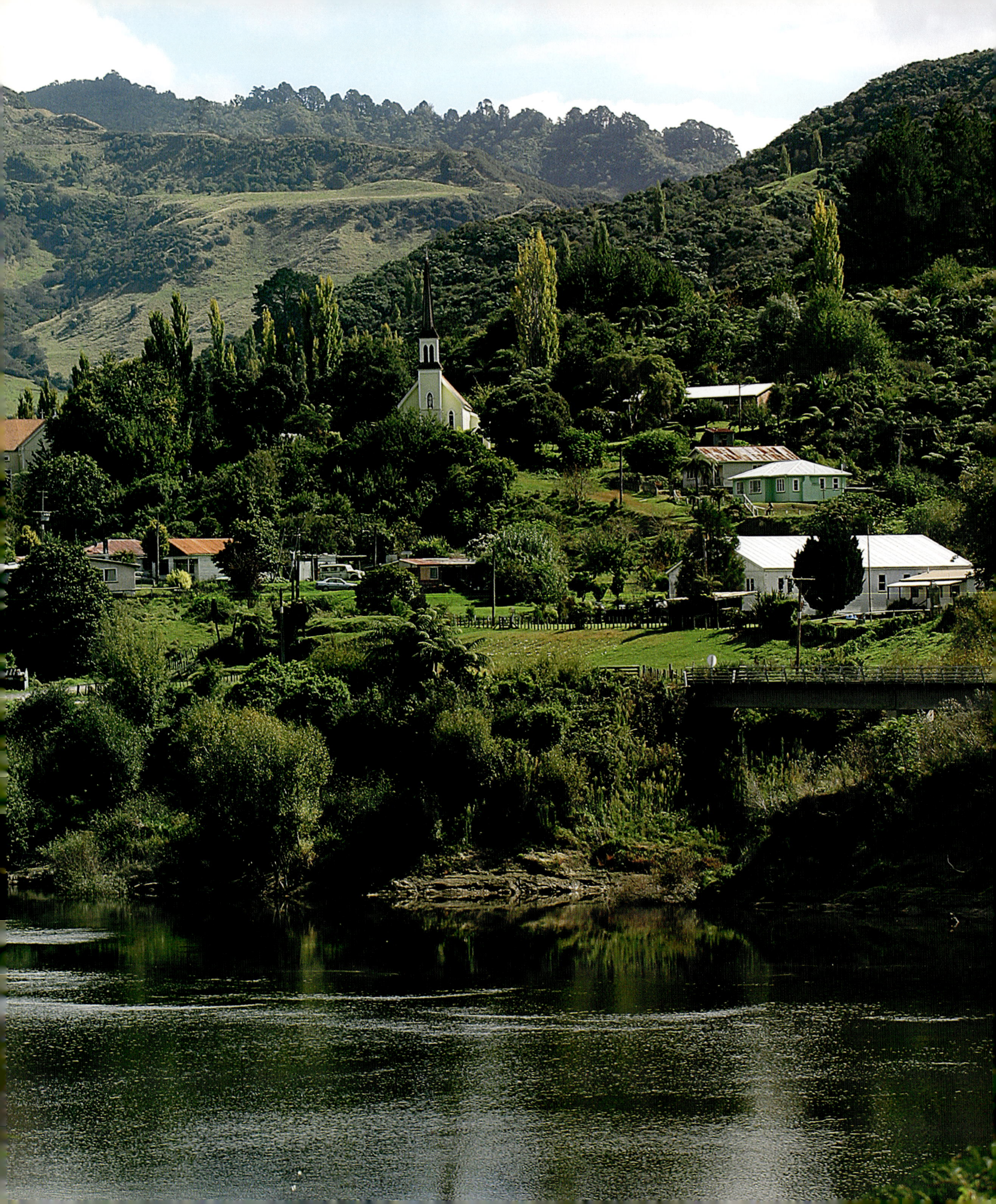

WHANGANUI — ANCIENT RIVER ROAD

Try to follow the Whanganui River on a map and chances are you'll lose it several times before you trace its beginnings to a small stream on the flanks of Mount Tongariro. It's a meandering river that travels north and east before finally turning to the south. Many of its tributaries, including the Whakapapa, which meets the Whanganui at the tranquil village of Kakahi, begin in the mountains of the Central Plateau.

Some, such as the Ongarue, travel south from the forest area west of Lake Taupo, while the Tangarakau and the Whangamomona feed it from deep gorges on its eastern boundaries.

It's also a chameleon river that changes character almost as often as it changes direction. Initially a clear mountain stream gurgling its way across the volcanic plateau, it arrives in Taumarunui as a smoothly flowing stretch of water before it twists and turns its way south, a moss-green snake hindered by rapids and pushing its way through dense forest and awe-inspiring gorges before it arrives at Pipiriki.

From there it widens and continues its slow slide to the sea through bush-covered hills, broken here and there by the occasional small settlement. Through the city of Wanganui, exotic trees and suburban houses line its banks. It cuts a wide, glittering sweep before it curves out to the ocean alongside sandy dunes where once there were Maori fishing villages and later a whaling station.

A gloomy river in places, it is secretive and mysterious, only revealing its true character to those who know it well.

According to Maori legend, the Whanganui was born of seduction and violence, the fall-out from a love triangle. Mount Taranaki and Mount Tongariro, at the time close neighbours, both fell madly in love with beautiful Mount Pihanga. They fought over her and Tongariro won. Taranaki ripped himself out of the ground and fled towards the sunset, gouging a deep gash across the country as he went. Arriving at the coast, he turned north and settled where he remains in splendid isolation today. Soon after his flight, a spring of clear water rose from the side of Mount Tongariro and flowed through the chasm left in Taranaki's wake – forever after known as the Whanganui.

Kupe is credited with discovering the river and, much later, descendants of the *Aotea* canoe settled alongside the river, attracted by its abundant food supply – eels, native trout, lamprey, koura (freshwater crayfish) and black flounder. In the surrounding forest were birds such as kereru or native pigeon and other native birds that still inhabit the lowland forest, as do brown kiwi, whose call can sometimes be heard at night.

The Whanganui is the longest navigable river in the country. For centuries before the arrival of Europeans, Maori in their waka travelled the river's length as we travel a modern highway and its tributaries were side streets that led to the hinterland. From as far away as Kapiti Island, Maori paddled up the Whanganui to its junction with the Manganui a Te Ao, and from there drove their waka up against the current along a complicated network of waterways to Lake Taupo. Various tribes dominated different stretches of the river. Raiding parties from the volcanic plateau would travel down it as far as the coast, and hapu (sub-tribes) from the lower river would travel north to engage with allies.

Much more than a thoroughfare, though, the Whanganui nourishes the land; it succours the people who inhabit its shores and is an integral part of their mythology. Each rapid – and there are 239 – had a name and every significant bend of the river a controlling guardian spirit. Taniwha live in its depths. People have cultivated its terraces, taken food from its waters and drawn spiritual sustenance from it for 40 generations, and many today still have an almost

OPPOSITE: The Whanganui River town of Jerusalem, site of an early Catholic mission and, more recently, home to poet James K Baxter.

BELOW: The headwaters of the Whanganui River on the flanks of Mount Tongariro.

RIVERS OF THE WEST

TOP: Holes made over the centuries by Maori poling their river craft up the Whanganui.

ABOVE: Final resting place of the MV *Ongarua* at Pipiriki, one of the boats that plied the river in the early 20th century.

mystical relationship with the river. For Whanganui Maori the river will always be an indivisible ingredient of their ancestry.

It is because of this that Maori have fought tenaciously to have their rights to the river and its riches recognised and affirmed under the Treaty of Waitangi. Since the early years of the 20th century, there has been a continuing stream of petitions from Maori to government claiming compensation for the loss of land sold adjacent to the river; for loss of customary rights over the river; and, since 1937, for title to the riverbed itself, a claim that is still not settled. When the Whanganui National Park was set up in 1986, the river was expressly excluded from it because of this unresolved claim. The occupation of Moutoa Gardens by Maori in 1995 was the result of continuing frustration over lack of resolution of long-standing legal battles.

Newcomers

Whalers had come to the area previously but the first European settlers arrived in 1841. They purchased land from the New Zealand Company to start the town of Wanganui, so beginning a convoluted relationship with Maori, many of whom welcomed the settlers in spite of ongoing disputes over land ownership.

Due in large part to the arrival of missionaries, and particularly that of Richard Taylor, appointed to the Anglican mission in 1843, there was much early co-operation between the two races. Many Maori embraced Christianity. They adopted the cultivation of European crops, and several flour mills were set up between Wanganui and Pipiriki, which were of advantage to all those living by the river. Old fruit trees, willows, poplars and other exotic trees still mark places where early settlements existed.

During the first two decades of its existence, the town of Wanganui was threatened by Maori attack on several occasions. The battle of Moutoa Island in 1864 was a defining moment that left a legacy of mistrust between Maori and European. Essentially an offshoot of the war that broke out in Taranaki, it was a stand-off between the tribes of the lower Whanganui, who supported the settlers, and the more extreme tribes of the northern parts of the river, who were opposed to the alienation of any more Maori land.

A military post at Pipiriki defined it as the demarcation point between hostile and friendly Maori. Skirmishes continued into the 1880s, but by then the town of Wanganui was well established and the influence of settlers was changing life on the river, although opposition to European encroachment continued well into the 20th century, especially on the upper reaches of the river.

Another missionary who exerted a long-lasting influence on the Whanganui was Suzanne Aubert. She arrived in 1883 at Jerusalem (Hiruharama), and stayed for 16 years. She bought land and later set up the Sisters of Compassion, a religious order dedicated to helping the poor and children. She established a school at Jerusalem, planted an orchard, produced and sold medicines and published a book of Maori conversations, before she moved to Wellington.

Another famous resident was the poet James K Baxter, who established a commune at Jerusalem in the 1970s and was buried there in 1972. Today the settlement has shrunk in size, but three sisters of the order still maintain a presence there, and the historic St Joseph's Church, visible from the River Road, is one of the most photographed churches in the country. The old convent is available for group visits and retreats.

River traffic

Mention the Whanganui River now and most people think of canoeing and kayaking. In a curious reversal, this is how it always was. For centuries, waka were the mode of transport for Maori; later they provided essential services for early settlers on the river. In the latter years of the 19th century, when the Whanganui River was advertised in Britain and Europe as one of the great scenic attractions of the world, tourists would be taken on the river in 'waka tiwai', simple dug-out canoes carrying up to six people, which were easier to navigate than the bigger waka used for transporting equipment and provisions. It was not unusual though for some of the crew to get out and help pull the waka up rapids. A photo in the Wanganui Museum shows a canoe being poled up the river in 1885 with seven men on board, one very correctly clothed in hat, collar and tie.

By the 1880s, however, there were plans afoot to replace canoes with bigger vessels. Work had begun on building the Main Trunk Railway, and the river was seen as a route for transporting material north for its construction as well as providing a scenic link for passengers between the north and south railheads, before the whole line was completed in 1908.

A dependable, regular steamship service was up and running by 1891. It formed the basis of a business empire developed by Alex Hatrick, an entrepreneur of the times. Mayor of Wanganui from 1897 to 1904, he expanded his shipping business (which at its height included 12 steamers as well as several motorised canoes) to include buses, car sales, liquor sales and Pipiriki House, which he bought in 1899

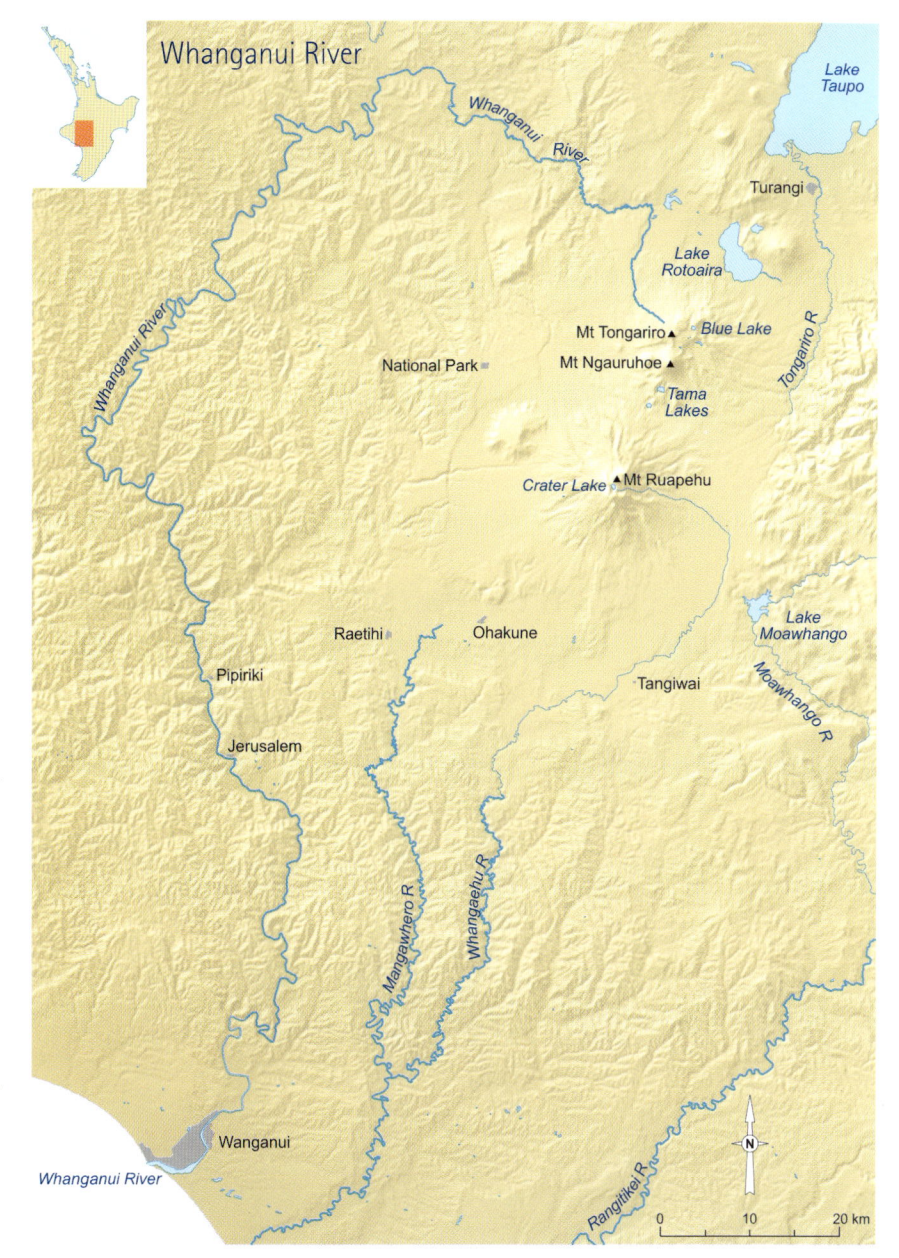

ACTIVITIES

CANOEING/KAYAKING: Hire or tour options (over 200 km).

CAMPING: Huts and campsites along the river.

WALKING AND TRAMPING: The Matemateaonga Walkway in the Whanganui National Park covers 42 km and starts inland from New Plymouth, following an old settlers' dray road and Maori trails to reach the Whanganui River. Five huts along the way.

The Mangapurua Track runs 40 km between Whakahoro and Mangapurua Landing on the river, taking three to four days.

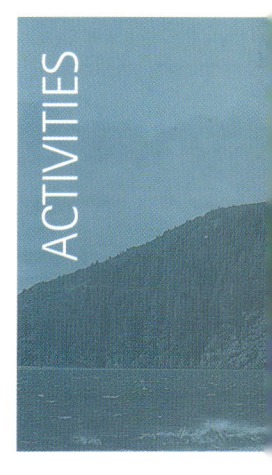

The city of Wanganui embraces both sides of the Whanganui River.

and completely rebuilt in 1903. When it burned to the ground in 1909, he rebuilt it once again. By the time it burnt down again in 1959, Hatrick had long been in his grave, the river trade was well over and all that can be seen now of the once fabled hotel is the place where it formerly stood.

Operating steamboats on the river was not without its complications. Big riverboats necessitated the modification of the river to make a safe passage possible. In 1891 the Whanganui River Trust was established with the aim of maintaining the river channel. It instituted the use of 'snagging punts' with crews employed to remove obstacles and generally keep the river open to boats. Not all the obstacles were minor. In 1903 a waterfall was blasted out of existence to create a more gradual, navigable slope, and a big rock weir was constructed to provide a channel for the boats. Sometimes boats had to be winched up shallows or rapids.

'Pa tuna' – traditional eel traps fashioned from manuka stakes and extending far out into the river to trap the eels on their journey back to the sea – caused endless problems. When they were dismantled to make way for the steamers, Maori protesters disrupted the work and rebuilt weirs that had been demolished. A treasured part of traditional life for Maori, their eventual removal laid the basis for continuing claims for compensation.

The presence of the snagging punts, which continued well into the 1930s, facilitated the arrival of more settlers along the Whanganui. As bush was cleared for farming, the run-off from the hills created frequent fluctuations in river levels.

At the beginning of the 20th century, concern about the clearing of land and subsequent loss of scenic areas promoted the establishment of a Scenery Preservation Commission. Its members travelled the Whanganui deciding on land near the river to be put aside as reserves, the creation of which cut across cultural boundaries and ancestral rights. The two races could agree neither on the use of the water nor on the use of the land bordering the river. Some reserves prevented access to the river by Maori. Such ownership problems caused conflict that persisted through most of the ensuing century.

The number of passengers on the river started to decrease after the completion of the Main Trunk Railway in 1908; this, along with the silting up of the river in the 1920s, led to the decline of the riverboat trade, and forestry took its place as the main economic activity. By the end of the 1950s, the last of the historic paddle-steamers had gone.

The city of Wanganui

The confusing difference in the spelling of the names for the town and the river is the result of historical changes. The area was originally called Whanganui by Maori, meaning 'big bay' or 'big harbour'. Possibly the influence of the local Maori dialect played a part in the town's name being spelt without an 'H', and the name for the river followed suit for a long period. However, the river's name reverted to Whanganui in 1991, although the name for the town did not change.

The fortunes of Wanganui have followed a similar pattern to the development of the river. A busy port during the 1920s, its importance declined in the next two decades due to competition and shipping centralisation, but in the last 20 years Wanganui has experienced something of a renaissance. With a population of nearly 40,000, the town has become a centre for the arts. The establishment of the Whanganui National Park has increased the number of visitors, eager to experience the beauty of the river and the surrounding bush areas. The Whanganui River, travelled by thousands of canoeists each year, is easily navigable for 170 kilometres through the park, though is not actually included in the park.

Along the River Road

The River Road links Wanganui to the tiny village of Pipiriki. A road not much travelled except by locals, who know every twist and each sharp bend, it follows the contours of the steep country high above the river. The views are fantastic – when you can take your eyes off the road. We catch a glimpse of an intriguing, blue-roofed house across the river. It offers accommodation evidently, if you're brave enough to travel over the water on a flying fox.

Here and there marae, many of which have been restored recently, are signposted along the way. Their names – Atene (Athens), Koriniti (Corinth), Ranana (London) and Hiruharama (Jerusalem) – are derived from suggestions made so long ago by the missionary Richard Taylor. With prior permission, travellers are welcome to visit some of them, and DOC camping grounds offer overnight facilities. Land here was balloted to soldiers returning from World War I, and attempts were made to open up the river valley and its tributaries as farmland. The river was the only means of reaching the farms. There's an area marked by old macrocarpas where the farmer used to lower his bales of wool over the cliff to a riverboat waiting below. Looking at the bony, eroded hills it's not difficult to see why the farms were not successful. Now tree ferns have spread thickly among the regenerating bush and rewarewa stands tall among it.

Pipiriki and most of the land down the river is in Maori ownership. There are no shops, no fuel available and the rural mail comes through five times a week. When we arrive at the village it shows no sign of its former glory when paddle-steamers disgorged tourists at the foot of a zigzagging set of steps that led up to Pipiriki House.

A view from the water

We explore the upper reaches of the river by jet boat, and it is not long before we are cruising through a canyon of perpendicular cliffs, thickly clad in bush. Our guide points out where the old riverboats had to be winched up the rapids. At Mangaio Stream he indicates the site of an ancient pa, high on top of the cliff. The only access was by a ladder, which could be pulled up if attack threatened. Further north, we pass the confluence of the Whanganui and the Manganui a Te Ao, now a sanctuary for the increasingly rare whio or blue duck.

Restored to her original splendour, the *Waimarie* paddle-steamer is a grand sight on the lower Whanganui River.

Kayakers stop for a break on the Whanganui River.

The 'Bridge to Nowhere' surrounded by bush — an example of the maxim 'timing is everything'.

At Mangapurua, deep in the Whanganui National Park, we pull into the landing and go ashore. A group of canoeists arrives barefoot. They've been on the river for four days, en route from Taumarunui. We all hike in to the famous Bridge to Nowhere, a monument to ill-conceived government planning of the 1930s. It was built to provide access to farm settlements which, even before the bridge was finished, were failing. Arching high above the Mangapurua Stream, the bridge is magnificent. So is the setting. Tree ferns in their hundreds successfully hide any traces of earlier attempts to tame this wild country.

Returning to the boat, we climb in and head back down the river. We pass beachheads of stranded driftwood waiting for the next flood to wash them closer to the sea. We peer in awe at neat round holes worn into the riverbank by those ancient canoeists poling their craft against the current. A heron flies silently overhead.

Is it too much to suppose that ghosts abide here, lurking just around the corner?

FLOOD WATERS FROM THE CENTRAL NORTH ISLAND

Sharply divided by the Ruahine Range and the Tararuas, the central North Island contains a mass of rivers and streams that rush down the sides of the ranges east and west before becoming entangled among the lesser hills that hinder their direct progress to the sea.

As forests have been felled and swamps drained to provide for agriculture, the loss of these natural sponges in an area of high rainfall has resulted in flood waters that have wrought enormous damage over the years, to towns as well as the land. In continuing efforts to minimise the floods' effects, the land itself has been much modified. But when heavy rain falls in the ranges, green fields where cattle graze in moderate weather can rapidly become wide expanses of swirling waters. Life for many farmers in this area is fraught with anxiety and frequent battles against the elements.

The Manawatu River

Unlike many other rivers in New Zealand, the Manawatu does not flow from a divide. It rises on the eastern side of the lower North Island, on the flanks of the Ruahine Range to the north of Norsewood, and starts its journey south towards Woodville, at which point it turns abruptly and heads westwards through the Manawatu Gorge, effectively a crack between the Ruahine and the Tararua ranges. Sediments found high above the gorge show that several million years ago this area was sea, with a strait forming a passage between islands, the tops of which are now the two ranges. As glaciation lowered the sea level the ridge emerged, as did a river draining the area previously occupied by sea. During the subsequent rise of the land, the river has cut a deep trench in the wedge of hard greywacke rock that has simultaneously been thrust upwards between two fault lines. Seeing the river as it powers through the gorge, you get the impression of a river in a hurry. Muddy more often than not, it is easy to imagine its scouring action as it grinds ever deeper into the rock that forms its base.

Where the river flows through the city of Palmerston North and continues on its way to the Tasman Sea by way of Foxton, it has a more manicured appearance. Sweeping curves of water spread through the Manawatu Estuary, creating an internationally known habitat for migratory birds, which assemble there during the summer months, among them godwits, red knots, the unique wrybill and the Pacific golden plover.

In early Maori times the route through the Manawatu Gorge was the only way to make the journey from one side of the ranges to the other. They called it Te Apiti, meaning the narrow passage, and the upstream journey by canoe involved much effort, paddling and poling the frail craft against the fast current. Near the centre of the gorge stands a large rock, reddish in colour, that held special significance for Maori. Named Te Ahu-a-Turanga, meaning a sacred place of Turanga, it was considered the guardian spirit of the gorge, and legend claims that the rock remains above water even during the fiercest flood.

The Manawatu River emerges from the Manawatu Gorge, a spectacular break in the Tararua Ranges, that run parallel to the west coast of the lower North Island.

RIVERS OF THE WEST

The swampy tidal flats of the Lower Manawatu River, near Foxton.

When Europeans arrived they also saw the river as the best means of transport, and it's no surprise that the most important town of the region, Palmerston North, is situated beside the Manawatu. When the town was founded in 1866, it was surrounded by forests, and all traffic, including felled timber, travelled by river to and from the then thriving port at Foxton. But a link was needed with Napier and the East Coast and in 1871 construction started on a road through the gorge. Initially a primitive route, it was wide enough only for one-way traffic. At the eastern entrance to the gorge, before a bridge was completed in 1875, passengers had to cross the river in a cage suspended 20 metres above the water. At the Ashhurst end, closest to Palmerston North, the crossing was made by a 'ferry' constructed of two Maori canoes roughly joined by a deck. Not surprisingly, the road was infrequently used until the second bridge was in place. Building a railway through the gorge was much more complicated and not completed until 1891.

As the river meanders across the coastal plain, its wide glittering stretches imply that this is a tame river, meek in character. On a good day this is true. But the Manawatu is fed by several major tributaries, all of which rise in areas where heavy rains are frequent. The Mangatainoka, Tiraumea and Mangahao rivers, all from east of the Ruahines, meet the main river between the town of Woodville and the Manawatu Gorge. At Ashhurst, on the western side of the gorge, it is joined by the Pohangina River and, near Rangiotu, by the Oroua, both of which rise in the western Ruahine Range in an area with an annual rainfall of about 500 centimetres.

As with so many other New Zealand rivers, the Manawatu has been transfigured by human activity. The forests that clothed its banks have gone, logged out by Scandinavian settlers in the 1870s. The western or seaward side of the Manawatu is a broad coastal zone that once consisted of sand dunes, covered with scrub and coarse grass, and swamps of flax and raupo, now drained to provide farmland. The sponges that helped soak up torrential rainfall before it entered the river and its tributaries have gone, and the Manawatu is now known for its rampaging and recurring floods.

A LAND OF WATER

Since the 1950s, enormous flood control work has been carried out across the Manawatu Plain. Stopbanks crisscross the land. Left to its own devices, the river would probably split into numerous braids. Now it is channelled and disciplined to follow a predetermined course – until the weather gods decide differently. In 2004 they decided to wreak havoc on the area. Shelley Dew-Hopkins, a farmer who lives in Opiki and who was involved in farm recovery operations with the rural sector after it was all over, talked about the floods that devastated the whole of the Manawatu-Wanganui district in February of that year.

'We're accustomed to floods once or twice a year. Even to having roads closed in the middle of summer. But this was totally different. I remember going to check on the river and I couldn't believe how fast it was rising. And how dirty the water was and how much debris was floating past. People in Palmerston North were shocked to see cows being swept along. We went up onto the stopbanks and all you could see was water right up to the edge – like a great tablecloth spread over the land. There was no way we could have walked the cattle out because of all the water. We were out sandbagging along with the army and the rest of the community. The river was frothing – it's the worst I've ever seen it.'

The flood banks that protected the Dew-Hopkins' farm held, but others lower down were not so lucky. 'The intensity was shocking,' Shelley continued. 'The smell of decaying vegetation and dead stock was awful. I'll never forget that. A terrible amount of rubble was left on paddocks – stones, boulders, rubbish and weeds. Farmers have weeds coming up now that they've never seen before. And the clean-up was made worse by the fact that there were more floods several months later.'

Luckily the stopbanks around Palmerston North had been increased in height during the previous five years. They were not breached and the city stayed dry. But 2004 is a year that will go down in Manawatu history for all the wrong reasons.

Not only notorious for its flooding, the Manawatu has also gained a reputation for high contamination caused by run-off from agricultural activity along it banks, not helped by a consent granted in 2006 to the giant Fonterra Dairy Company to discharge large amounts of 'waste water' into the river from its Longburn plant near Palmerston North.

Rangitikei

For some New Zealanders, the Rangitikei River conjures up an image from a Peter McIntyre painting of cool green-blue water and vertical cliffs, glaringly white in the sun. Keen rafters see it as a river of narrow gorges and challenging white water, while anglers regard it as a haven for a quiet, get-away-from-it-all weekend.

The course of the 240-kilometre Rangitikei is less complicated than many. It rises high in the Kaimanawa Ranges, not far, as the kereru flies, from

The aftermath of the 2004 floods that swept through the Manawatu. The Saddle Road Bridge was swept away (top), and the hamlet of Scotts Ferry (above) inundated by the flood waters.

The Rangitikei carves its way through the countryside, creating its distinctive cliffs.

Lake Taupo. It carves a turbulent course through the Mokai Gorge – an 80-metre-deep slot between towering cliffs hung with moss and ferns – and from there flows more or less south until it emerges from spectacular canyons and steep rapids to become a calmer, occasionally braided, river, on the large Manawatu Plain.

Like the Whanganui to the north and the Manawatu to the south, its catchment area, and that of its two main tributaries, the Moawhango and the Hautapu, includes a considerable chunk of the central North Island. When heavy rain falls west of the Ruahines, the Rangitikei, like its neighbours, is prone to disastrous flooding, due in part to the clearing of forest that has taken place along its borders. The massive flooding of 2004 caused a state of emergency to be declared in the Wanganui-Manawatu district, with several of the towns along the Rangitikei seriously affected. Residents in some areas had to be evacuated from their homes, and now, when you fly over the area, you see raw gashes slashed across the hillsides, evidence of long-term damage caused by the torrential rains.

When it was founded, the town of Bulls was briefly known as Rangitikei, but in 1872 it was changed to recognise James Bull, a prominent pioneer in the town. Along with Marton, and Taihape further north, Bulls is now the centre of an agricultural community, but the area of the Rangitikei is increasingly becoming known as a focus for adventure tourism and recreation. Visitors come to the area to fish, to explore the river in rafts or canoes, to enjoy exploring the many well-known gardens in the district and to absorb some local history in the course of visiting various old homesteads that were built in the late 19th century. The North Island's highest bungy jump is over the Rangitikei, not far from Taihape.

Because of the river's outstanding scenic characteristics, its opportunities for recreational fishing and its wildlife habitat, no hydro dams can be built upstream of Mangaweka, although there is a small power station nearby on the Mangawharaiki Stream.

And for those who like to collect statistics, the second-highest railway viaduct in New Zealand, at 81 metres, spans the Rangitikei between Mangaweka and Taihape on the North Island Main Trunk Line.

Early evening in the Rangitikei River Valley, near Mangaweka.

NELSON LAKES AND THE WEST COAST

NELSON LAKES

ABOVE: Early morning, Lake Rotoiti.

OPPOSITE TOP: Lake Rotoroa.

OPPOSITE BOTTOM: Fishing the Gowan River outlet at Lake Rotoroa.

Lakes take many different forms, depending on how they were originally made. Volcanic eruptions create lakes that are deep and roughly round in outline, the neighbouring hills often cone-shaped and rising sharply from the surrounding land. A lake ground out over time by slow glacial movement takes the form of a long trough, slotted into a valley at the base of continuous mountain ranges.

Lakes Rotoiti and Rotoroa at the hub of Nelson Lakes National Park are the latter kind, both formed by glaciation at the end of the last ice age. Mountains, clothed in beech forest that stops abruptly at the bush line, form a backdrop to the lakes. In winter their bare peaks, high above the lakes, are covered in snow.

Maori legend, however, credits the revered chief Raikaihautu with the creation of the lakes. On his exploratory trip through the South Island he gouged out these great troughs, as well as the deeper lakes further south. They filled with water and provided fish to feed his people in later generations. Midden sites uncovered at Kerr Bay on the shores of Lake Rotoiti and further inland at the head of the lake indicate that these were favoured stopping places for Maori on their journeys between Tasman Bay and Canterbury or the West Coast.

Europeans probably first became aware of the lakes' existence as a result of the voyaging of J S Cotterell in the early 1840s, when he explored the area with a Maori guide. From the south end of Lake Rotoiti, he followed the Travers River well into the interior and climbed a peak high in the St Arnaud Range, now known as Cotterell Peak. In 1846, artist Charles Heaphy, with his fellow explorers William Fox, Thomas Brunner and their Maori guide Kehu, crossed Lake Rotoroa in a canoe and continued on down the Gowan River on a voyage of discovery.

The first Europeans in the area were seeking land suitable for grazing sheep. Few fertile plains existed around the lakes – fortuitously as it happened, because this saved the region from the exploitation typical of the time. As early as 1907, some land was set aside for public use, but the actual National Park was not created until 1956. The small alpine village of St Arnaud serves as its gateway.

Although the park is a popular destination for trampers, hunters and fishers, who come to the region

A LAND OF WATER

to explore the headwaters of the numerous rivers and streams that drain into the lakes and eventually feed the mighty Buller River, these are tranquil lakes and have none of the tourist bustle associated with some of the better-known lakes further south. Beech forest hugs their shores and boat houses built beside Lake Rotoiti by early bach owners have been required by park management to be removed. We camped there one spring when the kowhai trees close to the lake were in full flower. In the early morning the bush was dark. Mist drifted above the water like an imperfectly hung stage curtain. Tui clacked and warbled in the still air. As the mist rose the mountains appeared, guardians of the lake.

In 1997, DOC initiated the Rotoiti Nature Recovery Project which aims to restore about 5000 hectares of beech forest and, having instituted a massive predator trapping programme, hopes to establish a 'mainland island' to protect the birds already there and reintroduce more species formerly endemic to the area, including the South Island tieke or saddleback, kakariki, kaka and the giant snail. A community group called the Friends of Rotoiti helps by checking traps.

Lake Rotoroa is a short drive from St Arnaud. The road leaves the Buller and follows the short but turbulent Gowan River before arriving at the lake. Fed by both the D'Urville River and the Sabine, which itself drains the smaller Lake Constance nestled in beech forest in the ranges south of Rotoroa, this lake is a favourite haunt for fly fishers seeking brown trout. The day we visit, a lone fisherman is trying his luck at the outlet to the Gowan. It's a sparkling day with snow visible on the mountain ridges. Tui are voluble in their appreciation, but there is a sinister note in paradise. A Biosecurity officer patrols the lake edge, warning any boatie who arrives of the perils of didymo, and other lake and river weeds. It's a solemn reminder of the fragility of our waterways.

THE BULLER AND ITS TRIBUTARIES

The rivers of the South Island are usually in a hurry. Often steep and fast-flowing, they rush from the mountains to the sea by the most direct route they can find. Such a river is the Buller – known in Maori as

ACTIVITIES

MOUNTAINEERING: Good winter climbing routes, suitable for experienced trampers and climbers.

SNOW SPORTS: At Mt Robert, a small club field above Lake Rotoiti, and Rainbow, a commercial field in the St Arnaud Range.

WALKING AND TRAMPING: Most popular is the Travers/Sabine Circuit and the trip up to Lake Angelus, a beautiful alpine lake in the Travers Range.

WATER-SKIING: On Rotoiti but not Rotoroa. Mooring is not permitted.

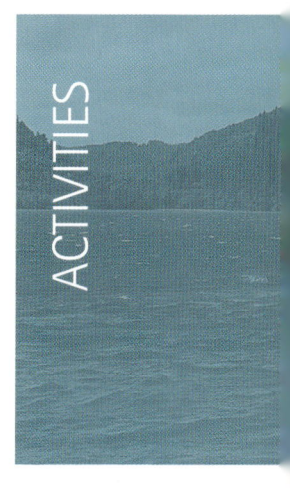

A quiet stretch of the Buller River.

Rust-encrusted rocks on the Gowan River.

Kawatiri, believed to mean deep and swift.

The largest river on the West Coast, the Buller is a highway, 169 kilometres long, that flows from its source at Lake Rotoiti to Westport and the Tasman Sea. Once used by Maori in their trading and transport of greenstone, its course now marks the beginning of the route followed by both the railway line and the Lewis Pass Road between the West Coast and Canterbury. Travellers going from Nelson to Westport meet the river at Kawatiri, and those from Blenheim, at Lake Rotoiti.

Marked by rocky gorges, inky dark bush, emerald willows, white-water rapids and sweeping expanses of wild water, it is joined by a mass of smaller waterways as it cuts its way westwards, creating a river with a larger flood flow than any other in New Zealand. Its main tributaries rise in the Spenser Mountains, in the ranges deep in the heart of the Nelson Lakes National Park. One such is the scenic Gowan, that provides a wild ride downstream for rafters serious about their sport and a pathway upstream for long-finned eels, eager to reach Lake Rotoroa.

Longer than the Gowan and fed by the considerable Glenroy River, is the Matakitaki. It starts life in the vicinity of the Lewis Pass and flows through steep mountain country before opening out into grassed river flats, finally joining the Buller at Murchison, as do both the smaller Matiri and Mangles rivers. Yet other tributaries, such as the Hope River and the Owen River, flow south to the Buller from the Kahurangi National Park.

The Buller River and the gorge it traverses have a formidable reputation. The scene of earthquakes, landslides and floods, the area has been a challenge to anyone seeking to explore its depths or tame its wilds. Horsemen, coach-drivers, road-makers and bridge builders of the past and truck-drivers, rafters and kayakers of today – all have tasted its dangers. But to drive the road beside the Buller and follow its turbulent, changing course west is an introduction to the somewhat mystical grip that the West Coast exerts over those who love its sombre colours and wild shores. During 1847, Thomas Brunner, accompanied again by the Maori guide Kehu, made his third journey along the Buller. They ventured into unknown country west of the town of Murchison and their journey from Nelson to the mouth of the Buller took them six rigorous months. Plagued by rain and hindered by the dense beech forest where no food existed, they almost died of starvation.

More than a decade later, in 1862, the discovery of gold nuggets in Lyell Creek by a group of Maori prospectors brought men to the district desperate to find riches. It wasn't long before Lyell, the usual gold-seekers' town, sprang up on the banks of the Buller. The following year, gold was discovered in the Mangles, Matakitaki, Glenroy and Maruia rivers, and the rush was on.

Later, with the discovery near Lyell of a quartz reef rich in gold, a stamper battery was set up, and between 1883 and 1900 several gold dredges operated on the Buller River between Lyell and the small town of Berlins, downriver from Inangahua. But the gold rush was short-lived, and when the precious metal became scarce, many men stayed on to work in the coal mining or timber milling industries. Others became involved in farming.

Chaos hit the small town of Murchison on the Buller River on the morning of 17 June 1929, when an earthquake measuring 7.8 on the Richter scale shattered the town. Though the earthquake was felt throughout the North and South Islands, it took more than 24 hours, in those days of limited communications, before the rest of the country discovered the extent of the disaster. In all, 17 people in the town and outlying districts were killed. Men working on a diversion tunnel aimed at enabling the Buller River bed to be mined were luckier. When a slip blocked access to the cage that normally took them across the river, they were left stranded, surrounded by water. Then the river stopped flowing altogether – dammed for 11 hours by an enormous landslide – and they were able to walk to safety. Another landslide blocked the Matakitaki River, forming a lake and raising fears of flooding when the dam broke. Roads, bridges and buildings were destroyed. Damage to the Buller Gorge Road was considerable and it took months to reopen the road between Westport and Reefton. The Maruia River jumped its banks and formed a new course. Where the fault line crossed the river, the land was raised several metres by the earthquake and the Maruia Falls were formed.

Another severe earthquake in 1968 affected Inangahua, a small town situated at the confluence of the Buller and the Inangahua rivers. So numerous were the aftershocks and so great was the damage to the majority of the houses that the whole town was evacuated. Fifty bridges were destroyed, the Buller River was temporarily blocked by landslides and 100 kilometres of railway track were deformed so seriously that it had to be replaced. Nature has a way of throwing nasty surprises at this part of the country.

Westport developed as an important town and port for the Buller District in the days when roads were few and first gold and then coal had to be exported. Ravaged by intermittent floods and affected in the latter half of the 20th century by the downturn in both coal mining and timber milling, the town has not always offered an easy life. But 'Coasters' are known for being a hospitable, resilient species, and in recent years the town has gained a reputation as a welcoming tourist centre. An open-cast coal mine operates nearby, and the port, kept open by regular dredging operations, offers cargo-handling facilities and a haven for fishing vessels. Whitebaiting is an important feature of local life.

With such a tumultuous flow, it is perhaps surprising that the Buller has not been exploited for hydroelectric power generation. Since 1904, various schemes have been proposed, but the unstable nature of the land and the risk of earthquakes have discouraged development.

With the decline of coal mining, gold seeking and timber milling, the Buller River and its tributaries are now the focus of recreational users. Born originally of gold fever, Murchison, on the banks of the river,

The Gowan River, a tributary of the Buller River, flows out of Lake Rotoroa.

is a farming settlement and has become a centre for outdoors enthusiasts. A paradise for fishing, rafting and kayaking, it also attracts jet-boaters and tourists eager to try their hand at gold-panning. The town lies at the southern gateway to Kahurangi National Park, while Nelson Lakes National Park is not far eastwards.

The Buller itself offers plenty of challenges for experienced 'river runners'. Although the best kayaking rapids are downstream from Murchison, those wanting to experience all its moods can put their canoes into the water at Lake Rotoiti and follow the river's course all the way to its mouth, just beyond Westport. For those with the courage, the daring and the skill, there is the challenge of the 11-metre Maruia Falls, which were first successfully kayaked in 1984.

WEST COAST RIVERS AND LAKES

From Karamea in the north to Dusky Sound in the south, the West Coast of the South Island is a world dominated by rain and rivers. Rain nurtures the dense forest that lends the land its wilderness aspect. Innumerable rivers, frozen or fast-flowing, carve out the landscape on their tumultuous journey from the Southern Alps to the Tasman Sea.

The rivers range in size from the mighty Buller to narrow, rocky mountain creeks. Some, like the Upper Karamea and the Fox, are characterised by house-sized boulders and awesome limestone cliffs. A few, like the Whataroa and the Wanganui, are short and meandering. Big or small, all are prone to sudden flooding. They carry heavy loads of gravel and sediments, which they deposit finally on the coastal beaches and in the ocean. Not surprising then that the harbours at Greymouth and Westport require constant dredging to keep them clear for shipping. Travellers following the length of the West Coast and crossing a constant succession of bridges can't help but be aware of the challenge presented to the engineers who first designed and built the road.

If anyone were asked to prepare a colour palette for the West Coast, the colours of the rivers would add a multitude of shades: peaty brown, soapy aquamarine, moss green, clear emerald. Add to those the bright red ochre of the lichen that clings to many of the boulders in the riverbeds, the patchwork greens of the bush, the changing shades of the sea that the main route follows and a picture is forming of this unique area of New Zealand.

The musical Maori names of the rivers add another dimension: Karamea, Mokihinui, Arawhata, Waimangaroa, Waianiwaniwa, Taramakau. Then there are the sober names of explorers who beat their way through untracked, often sodden, bush to chart this wilderness and who are remembered by the places they put on the map – Brunner and his lake, Fox and his river, Browning and his pass – giving them new names that displaced the Maori ones.

Before the arrival of the Europeans, generations of Maori travelled to the West Coast. They followed rivers on the east that led them to passes over the Main Divide. Once through the mountains, yet more rivers led them down to the coast. The greenstone or pounamu they sought on the West Coast was a precious trading commodity, prized throughout the country for the making of weapons and ornaments. One route followed the Hurunui through Canterbury to Lake Sumner, over what is now called Harper Pass, and down the Taramakau. Numerous lakes on this route would have ensured a supply of food as they travelled. This pass, in 1857, was where the first European crossed over the Main Divide.

Lewis Pass, also a much-used jade trading trail takes its modern name from Henry Lewis, a surveyor who, in 1860, was one of the first Europeans to travel that route. When gold was discovered a few years later, finding access through the Southern Alps to enable the transporting of gold to Canterbury became an imperative. Maori, who knew of the routes, were employed as guides by European surveyors. One, John Browning, gave his name to a range of mountains and the pass that Maori knew as Noti Raureka. Originally an important greenstone trail and now a popular route with trampers, Browning Pass is gained from the Wilberforce in Canterbury and leads into the Arahura River on the west. Gold, won from sluicing in the Arahura Valley, was taken into Canterbury by this track until the road was built over Arthur's Pass in 1865.

The riches that draw men to the West Coast have never been won easily. The gathering of greenstone had its dangers. Witness the grisly episode in the Lewis Pass area in pre-European times. When a group

Detail of the cable anchors on the Cook River Bridge.

A LAND OF WATER

The Cook River, one of the tumbling rivers crossed by the main West Coast highway.

of Ngati Tumatakokiri attempted to muscle in on the pounamu trade, a dispute is believed to have ended with a party of the North Islanders being ambushed at the head of the Maruia River. All were killed by Ngai Tahu and later eaten, giving the name to the area of Cannibal Gorge or Kopi o Kaitangata (a good feed of human flesh).

Gold extraction took its share of lives in the early days of mining and even occasionally in accidents through the 20th century as gold dredging continued with ever-bigger dredges. The coal mine disaster in 1896 at Brunner, on the north side of the Grey River near Greymouth, also took a horrific toll.

Nowadays the rivers and lakes bring visitors to the West Coast who want to kayak and raft their waters, try their hand at fishing or enjoy hiking the tracks along their borders. And after a strenuous day in the outdoors, tired visitors can relax in natural mineral hot springs offered by the Maruia Springs Thermal Resort. It is difficult to imagine a more idyllic setting than these pools, nestled on the banks of the Maruia River in a sheltered valley in Lewis Pass, with steam drifting quietly off into the surrounding bush.

On the Maruia also, good fishing is to be found a few kilometres down the road and downstream from Springs Junction. Around here, the river is swelled by several important tributaries including the Rahu, Woolley, Rappahannock and Warwick streams, all serious trout rivers in their own right. Anglers fishing the Matakitaki River need to be prepared to walk but it is apparently worth the effort, especially above Matakitaki Station. The Upper Grey and Inangahua Rivers are both known for large trout. The Grey rises in Lake Cristabel as the Blue Grey River and is joined by the Brown Grey a few kilometres further down. East of Lewis Pass, two rivers that yield trophy trout the most often are the Boyle and the Hope.

Patterns in the limestone rock at Fox River.

The distinctive 'S' bend of the railway bridge across the Grey River at Greymouth.

From September to the middle of November, the rivers of the West Coast have become a legendary destination for whitebaiters.

The Fox River rises in the Paparoa Range and flows out to the Tasman Sea, north of Punakaiki. A well-signposted walk from the main highway leads into an awe-inspiring cave on the riverbank or takes you further inland on an old pack trail. Short, like many of the rivers in this area and shallow near the coast, its water is stunningly clear and flows through beautiful West Coast scenery. Vertical limestone cliffs rise above the river with deep green pools at their feet. Great chunks of rock that have broken away from the cliffs create islets in the water and nikau palms in profusion stud the bush. Walking along its banks takes you to the kind of place where you can sit and dream, and immerse yourself in the life of the bush.

Grey River

One of the larger rivers on the coast, the Grey rises at the Main Divide, as does its largest tributary, the Ahaura. Maori called Grey River Mawhera, and the pa at its mouth, where the town of Greymouth now stands, is reputed to have once been a centre for the working of greenstone. Nowadays, workshops displaying greenstone carvings and jewellery are part of the tourist industry that attracts visitors here.

In the course of his journeying in 1846 and 1847, Thomas Brunner discovered a seam of coal on the north side of the Grey River. Later, coal became a mainstay industry of the area and was barged down the river to its mouth from 1864 until the railway was completed 12 years later. Gold, from both alluvial recovery and dredging, was also important for a time. Tailings along the riverbanks are a legacy from the gold rush days.

Two notable bridges over the Grey talk of the past. Near the centre of the town is the old S-shaped railway bridge, built between 1896 and 1897, the only one of its kind in New Zealand and recently replaced by a new bridge nearby. Further upstream at Brunner is an old suspension bridge with a somewhat chequered history. Originally built to take coal from the mines on the north bank of the river to the railway on the south side, it actually collapsed before it was completed – luckily in the wee small hours of a July morning, so no one was killed. Rescued and completed, it first carried coal wagons, was later converted to accommodate horse and vehicular traffic, and is now solely a pedestrian bridge. Over the course of its life it has fallen into disrepair on several occasions and was finally rescued from destruction in recent years when a coalition of groups decided it was of historic importance and needed to be repaired and saved for posterity.

The Upper Grey is renowned as a white-water rafting river where adventurers enjoy its scenic gorges and backdrop of dramatic mountains. Anglers visit it too, to enjoy its bounty. More recently, scientists have been investigating the population of glass eels in the Grey, with a view to farming them as a food resource.

Lake Brunner/Moana and the Arnold River

A short distance from Greymouth, Lake Brunner is an idyllic destination for city-dwellers seeking the peace and superb fishing long enjoyed by permanent residents, who also used to enjoy an unspoiled view of their lake and the mountains in the distance. Unbelievably, and above the objections of the residents, recent permission has allowed houses to be built at the lake edge, directly in front of the small village of Moana, the heart of this settlement for a long time.

The Arnold River, moody, brown and slow-moving at its source at the northern end of the lake, is lined with thick bush and tall kahikatea. The harsh cry of paradise ducks follows us as we cycle along a track bordering its peaty waters.

A LAND OF WATER

The Arnold River outlet at the northern end of Lake Brunner.

Maori used to come up the Arnold to get seasonal food supplies at the lake. The river's Maori name of Kotuku Whakaoho commemorates an incident in Maori lore when Ngai Tahu came up the river planning to attack the local tribe. The locals got wind of their advance and planned a pre-emptive strike. But a white heron, disturbed by their manoeuvrings, took off with a great flapping of wings, the noise alerted Ngai Tahu and they emerged from the skirmish as the conquerors.

A small hydroelectric power station exists midway down its length, and below this the river attracts kayakers and rafters.

Taramakau River

One of the main sources of greenstone for Maori in days gone by, the Taramakau River starts life high in the mountains at the edge of Arthur's Pass National Park and as it nears the coast it opens out into a wide gravelly riverbed.

'When bank to bank with water, the Taramakau carries a huge burden of rock, shingle and boulders which make the riverbed a thunderous bowling alley of big goolies crashing together as they tumble along,' writes David Young in his book *Faces of the River*.

He also discusses at length the legal battles that a dairy farmer on the Hohonu (or Greenstone) River has had with dredging companies on the Taramakau over the years. With its neighbour the Arahura, the Taramakau had the doubtful honour of being declared a Sludge Channel under the 1886 Mining Act, which meant that any mining – and later dredging – detritus could be diverted into the designated river. Dredging accelerated in the 1930s, and as dredges became more effective, so it became more profitable. The last gold dredge on the Taramakau finally ceased operation in 1982. The result is a river of several braids and wide gravel banks, marked by continuing change of flow and channel as waste water and tailings have been churned around in it.

Happily whitebait have not been deterred from coming into the river and any day in the season a few hopeful fishers can be seen from the main highway, quietly working the current.

The span of the bridge and wide gravel bed hint at the power of the Taramakau River in flood. Dubbed the 'terrible cow' by early settlers, the river was notoriously dangerous and many lives were lost in its crossing.

NELSON LAKES AND THE WEST COAST

ABOVE: Lake Rotoroa, Nelson Lakes National Park.

OPPOSITE TOP: The sweep of the Buller River as it squeezes through the Buller Gorge.

OPPOSITE BOTTOM: Where the Wanganui River meets the sea — Wanganui Bluff at Hari Hari.

A LAND OF WATER

NELSON LAKES AND THE WEST COAST   89

Wild waters of the Haast River at 'Gates of Haast'.

Early morning reflections, Lake Matheson.

ABOVE: Kahikatea magic in a corner of Lake Brunner.

RIGHT: A leisurely cruise on Lake Brunner.

A LAND OF WATER

The secluded upper reaches of the Fox River can only be reached by a walking track that includes several river crossings.

The Maruia Falls have only been in existence since 1929, when they were created by earthquake activity.

RIVERS OF THE EAST COAST

RIVERS OF NORTH CANTERBURY AND MARLBOROUGH

Late spring brings a spectacular show of California poppies on a tributary of the Wairau River.

A flight over the top half of the South Island helps visitors to appreciate the topography of this rugged, isolated part of New Zealand, where beech forest once covered many of the hills and side valleys. In inaccessible places remnants of this forest still exist, and the trees take on a bluish tinge in low light.

Scree slopes give way to wide tussock lands along the valleys, and three major rivers tumble and twist their tortuous ways to the ocean. The Wairau, Awatere and Clarence follow roughly parallel routes as they travel generally northeast, though the first two are widely separated. Settlements have little impact on the landscape, and life on the few privately owned sheep stations in the area requires a high degree of self-sufficiency.

It is here that Molesworth Station is found. The largest sheep station in the country, it is a vast expanse of tussock and grassland, picturesquely set between the Inland Kaikouras and the Spenser Mountains. Farming in this dry climate, buffeted by northwesterly winds, has been harsh on the land, and considerable erosion has occurred over the years. Molesworth is now owned by the government and administered by DOC. It will be interesting to see whether improvements can be effected in the future. When conditions allow, Molesworth can be explored by road, although some parts are accessible only by four-wheel-drive vehicles. Two routes lead through the station: the Acheron Road from Blenheim to Hanmer Springs follows first the Awatere River and then the Acheron until its confluence with the Clarence; the Hanmer-Rainbow route, suitable only for four-wheel-drive vehicles, follows the course of the Clarence and the Wairau.

Wairau River

Starting life in the Spenser Mountains in Nelson Lakes National Park, near alpine Lake Tennyson, and fed by the Rainbow River, which rises in the St Arnaud Range, the Wairau is Marlborough's longest river. Its Maori name, meaning 'many waters', seems particularly appropriate, given that it is fed by numerous creeks as it flows north before changing direction and carving its way east through an area of glacial outwash. At this point it changes from a small mountain stream and frays out to become a typical New Zealand braided river, similar to those found in the central part of the eastern South Island.

In earlier times, the valley through which it flows was forested. When Europeans arrived in the area, large swamps existed in the lower plains, the north bank of the Wairau was covered with podocarp and broadleaf species, and the south side was tussock grassland. Cleared for farming, like so many other lowland areas in the country, the plains are now subject to severe flooding.

Nevertheless, the river is the lifeblood of a dry region. It provides the water that farmers depend on and is an essential ingredient of Marlborough's wine industry, which has burgeoned to become the largest wine production area in New Zealand. More than 16,000 hectares of land are planted in grapes. The river soils of the Wairau Plain are typically silty and stony, ideal for production of sauvignon blanc, a variety that has made Marlborough's name in the wine world.

Wild and scenic in its upper reaches, the Wairau is, at the time of writing, one of the last unobstructed braided rivers. The endangered black-fronted tern lives here, and banded dotterel, the threatened black-billed gull and Caspian terns come here to build nests at the end of winter and raise their chicks. It is also home to a diverse, native, braided-river fish population and attracts keen anglers to its waters.

Once known throughout the country because of the battle that took place on the banks of the Wairau between Maori and European settlers in 1843, the river is once again a site of dissension and distrust. A proposal to build a hydro scheme on the river, which would also make provision for irrigation, has aroused the opposition of DOC, Fish and Game, and The Royal Forest and Bird Protection Society and given rise to a local group determined to preserve the present form of the river.

The group called 'Save the Wairau' argues that the scheme will reduce the 49-kilometre middle reach of the river to a 'virtual trickle' for much of the year, ruining one of the world's few remaining, largely untouched, braided rivers and irrevocably harming the wildlife habitat.

Where the Wairau opens out into the ocean at Cloudy Bay, not far from Blenheim, there is an area covering about 2000 hectares of saline marsh and mud flats. Ecologically and archeologically these lagoons are significant. Big Lagoon is separated from the ocean by Boulder Bank. In times gone by, this must have been an important gathering place for Maori, for archaeological sites have been excavated exposing middens, campsites and burial grounds. Moa skeletons have been recovered from these sites. Moa hunters evidently herded the giant birds from the Wairau Plain and surrounding hills into this trap, where they were slaughtered.

Seen from Molesworth Station Road, this stretch of the Clarence River, at its junction with the Acheron River, is accessed after leaving Hanmer Springs and crossing Jacks Pass.

Before the arrival of Europeans, this area was a large body of fresh water. It was subsequently drained and turned into fertile farmland. Now treated waste from the Blenheim oxidation ponds is discharged through a series of man-made wetlands, no deeper than one metre in most parts, which host a wide variety of wildlife, including a large colony of royal spoonbills. Fish species, including galaxids, bullies and herrings, are common as well as larger fish such as kahawai and barracouta. A feature of the lagoons is the wreck of the *Waverly*, which was towed here and dumped as a flood control measure as well as a practice target for the army.

Awatere River

The name Awatere means 'fast-flowing stream'. The river rises far inland and flows along a splinter of the alpine fault between the Rachel Range and the Inland Kaikoura Range. In its lower reaches the riverbed opens out into dry tussock lands. Farming and more recently grape-growing in the region both depend heavily on the water provided by the Awatere. A long river it may be, but this is dry country, and as vineyard expansion has moved into the Awatere Valley, the need for a reliable water flow has become more urgent. Recently between 30 and 40 dams have been constructed in south Marlborough and more are planned, thus making provision for irrigators to draw water when natural flows are high, and cut back or shut off their takings when levels get low.

A gorgeous green pool on the Hope River at 'The Poplars'.

Most people driving State Highway 1 in Marlborough recognise the Awatere River by the double-decker combined road-rail bridge that crosses the river near Seddon. It was opened to traffic in 1902 and such is its exposure to wind that after the bridge was finished, extra money was needed to construct a windbreak for the top deck used by trains. Now, more than 100 years later, the timber deck is nearing the end of its life and a new bridge is planned for 2007. Needless to say, it won't be of double-decker construction!

Clarence River

The Clarence River, perhaps the most popular of these inland waterways among rafters and kayakers, is also one of the most convoluted. It drains from Lake Tennyson, high up near the Spenser Mountains in Nelson Lakes National Park, close to the beginnings of the Wairau. Unlike the Wairau, however, it heads south. At St James Station, not far from Hanmer Springs, it changes its mind about where it's going and dives off in an easterly direction before heading northeast between the Inland and the Seaward Kaikouras. At the end of the mountain range it makes another U-turn and goes south again to meet the Pacific north of Kaikoura.

This river offers an unforgettable white-water experience for rafters or kayakers, a three- or four-day wilderness journey from the mountains to the sea, passing through spectacular gorges and the open valleys of several high-country sheep stations. Boasting names like Jawbreaker, Carnage Corner and Sawtooth Gorge, there are rapids to try the skills of even experienced water rats.

Waiau River

The Waiau River rises in the Spenser Mountains and flows eastward to the Pacific Ocean. Two of its main tributaries are the Lewis and the Hope rivers. It offers the beauty of bush-clad gorges, fun rapids and lonely landscapes to anyone who enjoys fishing, rafting or kayaking. At the Waiau Ferry Bridge, jet-boating and bungy jumping are also available. In its upper reaches it is a big fish river. Access by vehicle is restricted to the private road through St James Station and it is a long trek in a 4WD vehicle from Hanmer Springs through the Clarence Valley and across Malings Pass to the upper Waiau. Bagging a trophy fish at the end of the day makes it all worthwhile.

The spectacular bridge over the Waiau River near Lochiel, on the road to Hanmer Springs. The first bridge across the Waiau was constructed in 1864, improving access to Hanmer Springs. Ten years later it was blown into the river and it was 1887 before a new bridge was in place. In the interim, many lives were lost trying to ford the river.

CANTERBURY'S BRAIDED RIVERS

High up in the Southern Alps is a collection of small lakes that few people ever see. Some of them don't even have a name. Mountaineers and keen trampers come here, and a few adventurous kayakers who are seeking the experience of a lifetime, paddling among icebergs. Icebergs? In New Zealand? Yes. For these are terminal lakes at the foot of glaciers where chunks of ice break off from time to time and float, captured there for a period by freezing temperatures.

Small they may be, but these lakes, and the glaciers that spawn them, play a continuing role in defining the central South Island, one of the most dramatic landscapes in New Zealand. For the thin streams that emerge from these chill grey waters bring with them cloudy particles of sediment, known as rock flour, that cause the surreal turquoise colours of Lake Tekapo and Lake Pukaki further downstream, which, along with Lake Ohau, were gouged out by the glaciers of successive ice ages that began about 200,000 years ago. The glaciers have receded a long way, leaving behind them massive moraines, huge scars and signature terraces on many mountainsides that mark the former extent of the glaciers.

These rivers of ice also gave birth to a series of braided rivers, which are a feature of the eastern side of the South Island – and few other places in the world. The poet Mary Ursula Bethell described these rivers in *By Burkes Pass* as

> ... *loose glacier-shed*
> *Fierce-laughing streams in circuitous riverbed.*

The principal examples of this distinctive form are the Waimakariri, the Rakaia, the Rangitata and the Waitaki. They in turn are fed by rivers that also form wide, braided beds. Within their boundaries lies some of the most spectacular wilderness scenery to be found – remote mountain ranges, beech forest, alpine herb fields, lakes and mountain tarns gouged out by glaciers, and breathtakingly beautiful river valleys.

OVERLEAF: The mighty Rakaia River, one of Canterbury's classic braided rivers, descends from its home in the mountains to cross the Canterbury Plains.

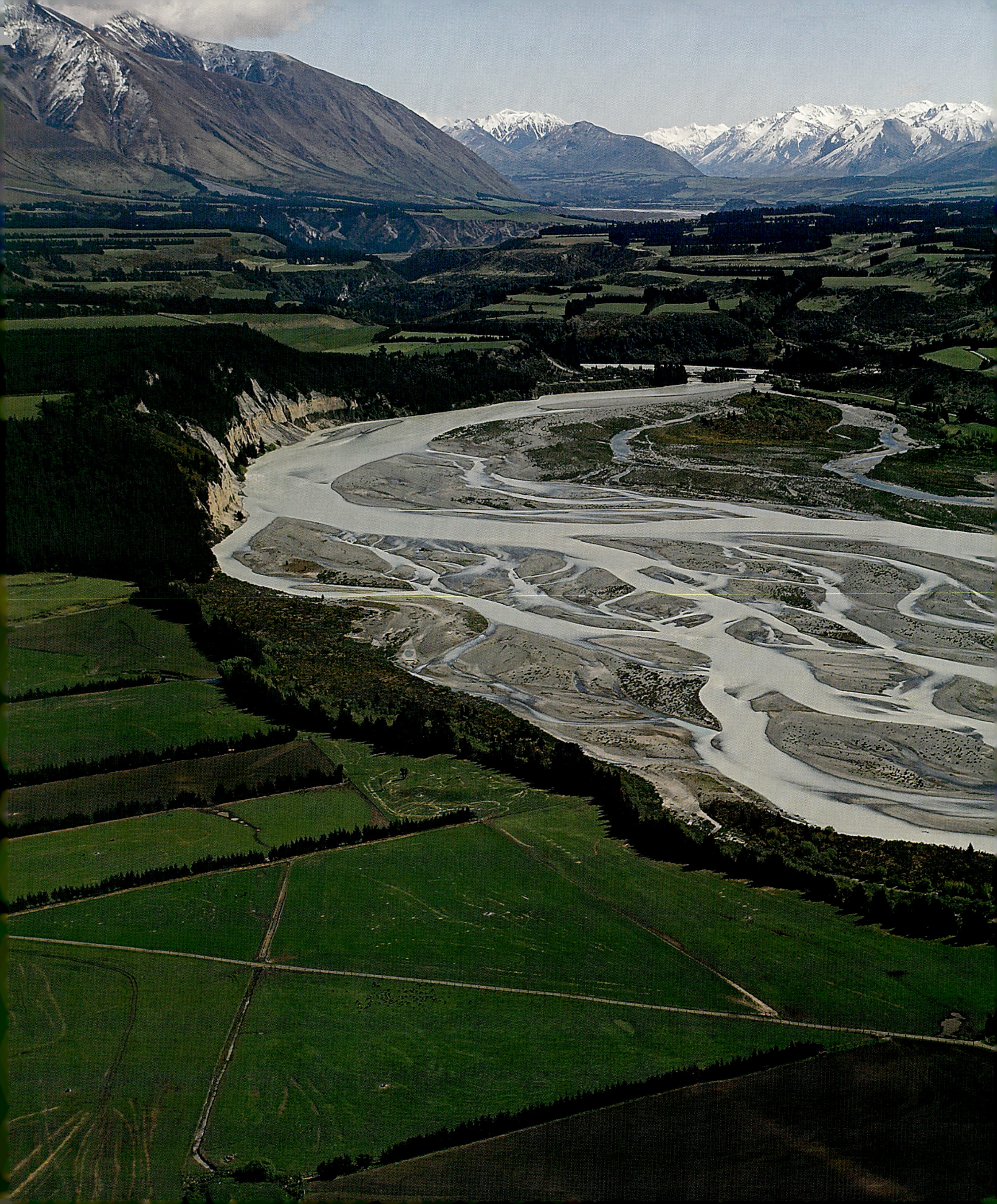

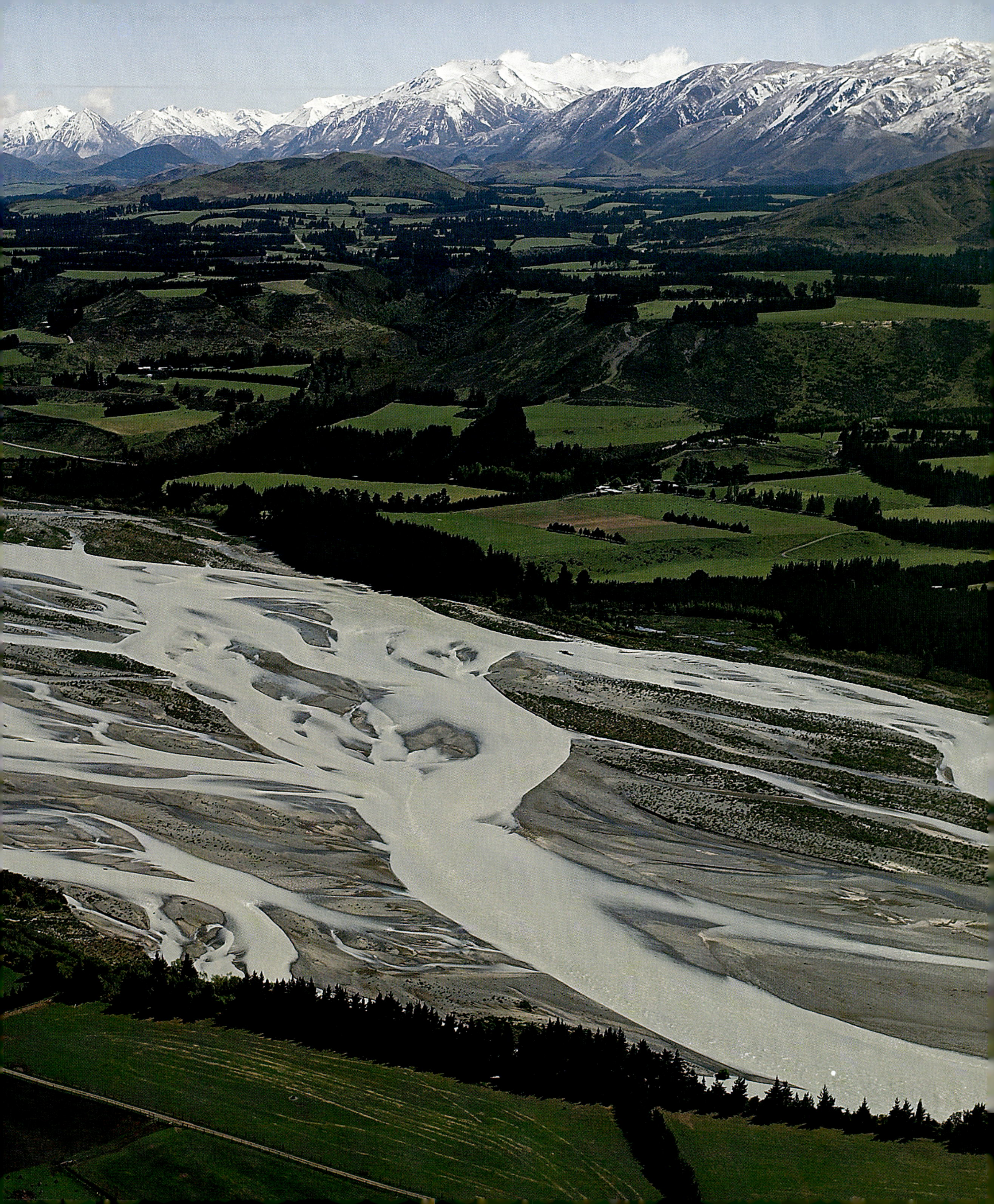

Boats moored on the Waimakariri River at Kaiapoi, north of Christchurch.

The shaping of the landscape and the rivers accelerated with the climate change at the end of the last ice age, about 10,000 to 12,000 years ago. As the glaciers started to melt they formed streams that found their way to the coast. Aided by heavy rainfall and constant erosion, the streams became rivers that brought down huge volumes of gravel, which built up in the wide valleys at the base of the mountain ranges. The process is continuous, with a constant movement of sediment, gravel and rocks as floods and erosion occur.

Often it is hard to trace the beginnings of these braided rivers, hidden as they are in neighbouring mountain ranges. Seen from the air they describe a woven fantasy in every shade of blue, grey, turquoise, mauve and green. As you follow their progress on a map, they evolve into a recognised pattern. Starting as thin streams, they mingle and grow and spread out into thin tentacles whenever the terrain gives them an opportunity. As they fan out towards the coast, where they feed into the Pacific Ocean at widely dispersed sites, their trail is one of wide expanses of gravel, divided into islands and restless streams meandering at the whim of the weather and the weeds that threaten to choke their natural wanderings. Insignificant during periods of drought, they fast become terrifyingly wide and turbulent after heavy storms in the mountains or warm nor'westers, which cause snow melt – with occasionally disastrous results. In mid-summer 2002 floods scoured out the approach to the rail bridge over the Rangitata – the longest rail bridge in the country. Two engines fell into the river, and the driver was lucky to escape with his life. The Main Trunk Line was closed for several days.

The northernmost of these rivers, and one of the largest, is the Waimakariri, emerging into the ocean at Kaiapoi, north of Christchurch. Downstream from its source among the untamed beauty of Arthur's Pass National Park, the Waimakariri is joined by its main tributaries, the Poulter, the Beasley and the Esk. Like its sister the Rakaia, this river also changes as it flows east, from deep fast water that plunges through narrow gorges, to woven braids of water flowing over wide gravel riverbeds. Given its size and proximity to Christchurch it is a flood hazard, and protection works have been carried out on the lower river. Some of its water is taken for irrigation and it is also used to provide for stock and domestic users.

The untamed upper reaches of the river are a favourite destination for jet-boating, rafting and kayaking, while the braided section attracts salmon- and trout-anglers. Residents of Christchurch regard it as their neighbourhood playground and it is the most used jet-boating river in the country. The lower part of the river is also an important habitat for wrybills, black-fronted terns and the banded dotterel.

One of the best examples in the world of a braided river, is the 145-kilometre-long Rakaia. Fed by melting snow and ice from the Lyell and Ramsay glaciers, the Rakaia is braided along a major section of its course. It drains a wide area of the Main Divide between the Rangitata and the Waimakariri, rugged country in the heartland of the Southern Alps. Upstream of the Rakaia Gorge, its wide gravelled bed has been built layer by layer as the surrounding mountains have eroded during successive periods of glaciation. It spills into the valleys at the base of the ranges, along the boundary between the Selwyn and Ashburton districts and out across the Canterbury Plains before joining the Pacific Ocean just south of Lake Ellesmere.

Its main tributaries, the Mathias and the Wilberforce, start high in the Southern Alps. They converge in a wide delta, and where the Wilberforce River meets the Rakaia is Mount Algidus Station, established in the 1860s and made famous by Mona Anderson in her books about life in the high country with its majestic mountains and graceful hanging valleys, and wide river flats. The eerie drama of the nor'wester builds here, to roar down the valley of the Rakaia, raising waves sometimes a metre high on nearby Lake Coleridge. Named by an early Canterbury settler, a relative of the English poet Samuel Taylor Coleridge, it was known by Maori as Whakamatua and favoured as a stopping place on their journeys to and from the West Coast because of its plentiful supply of eels and birds. The lake sits in a stark, ancient landscape. Lonely country this. It breeds self-sufficient people or it breaks them.

Once the mouth of the Rakaia was the site where Maori cooked moa, a huge, and now long extinct, native bird, which they captured further inland. Today, within easy weekend access of Christchurch, it is a popular destination for jet-boaters. The gorge area provides thrills for boaties, and anglers are tempted by the salmon and trout, which both thrive in its mountain-fed waters. As with the other braided rivers, the habitat it provides is important for several species of threatened water birds.

Look at the Rangitata on a map and its upper reaches are a tangle of multiple streams, fed by the Clyde and Havelock rivers, which, with their tributaries, drain the eastern side of the Southern Alps from the Lyell Glacier to the Godley Glacier. Other tributaries meet the Rangitata in a large basin, excavated by long-ago glaciers. Wild and scenic, surrounded by sweeping valleys and sharp mountain ranges, the Rangitata provided an ideal setting for filming some of the scenes for *Lord of the Rings*.

It also encompasses historic country. This is where the famous 19th-century novelist Samuel Butler came seeking unclaimed country in 1860 and eventually established his sheep station of Mesopotamia along the border of the Rangitata.

Although waters of the Rangitata are taken for irrigation, a Water Conservation Order, signed in 2002, caps that amount to one third of its flow and forbids the building of dams for hydroelectricity purposes.

The Waitaki

Perhaps the most complex of these rivers is the Waitaki. Not itself very long, its progenitors start far inland and spread over a wide catchment area, second

Picturesque Lake Coleridge perched above the Rakaia River. The difference in level between the lake and the river made Lake Coleridge an early site for a hydroelectric power scheme.

RIVERS OF THE EAST COAST

only to that of the Clutha. One of these, the Hopkins, starts life at the Richardson Glacier, at the head of Lake Ohau. It mingles with the Dobson River, which borders the eastern side of the Neumann Range, and together they enter the lake to emerge as the Ohau River. The Ohau continues by way of river and canal to find its way into the Waitaki.

Further east, the Hooker River, which begins at a terminal lake at the foot of the Hooker Glacier beside Mt Cook, and the Tasman River flowing from the Tasman Glacier, weave together across the wide gravel delta at the head of Lake Pukaki, and feed into the Waitaki, via Lake Benmore.

A similar course is followed by the Godley River, which rises in the terminal lakes of the Godley, Classen, Grey and Maud glaciers, wanders over a gravel bed in a series of thin streams at the head of Lake Tekapo and joins the Macauley to emerge as the Tekapo River, also finding its way into Lake Benmore. But with the construction of the three lakes, Benmore, Aviemore and Waitaki, much of the former character of the upper Waitaki River has been lost. Below the town of Kurow it unravels into many strands and meanders its way to the Pacific Ocean across a wide flood plain between Timaru and Oamaru.

The Waitaki and its tributaries have long been favoured as great fishing rivers. Brown trout were introduced as early as the 1870s and 1880s by acclimatisation societies already established in the area. These were followed by releases of rainbow trout, sockeye salmon and quinnat salmon in the early 1900s. The Waitaki River became a famous and popular salmon fishery, before the building of hydro power dams. Now landlocked sockeye salmon, brown trout and rainbow trout remain above the dams while quinnat salmon are restricted to the lower Waitaki River and its major tributary, the Hakataramea River. Below the Aviemore Dam, a spawning race one kilometre in length and capable of holding 3000 adult trout has been provided to maintain the trout fishery of Lake Waitaki.

The Waitaki River is a precious resource in an arid area, the water flowing in the river now a more valued asset than the gold that miners sought in rivers in the past. Competition for its use is heated. In 2003 the government passed special legislation to set up a statutory board whose job it is to look at the overall demands for water in the district and to devise a water allocation plan. With increasingly vociferous and often mutually exclusive demands for energy, irrigation and conservation, this promises to be no easy task.

The Ahuriri

Also feeding into the Waitaki is the fabled Ahuriri River – fabled because of its fishing, sometimes described as a 'unique angling experience'. Rising high in the ranges between lakes Ohau and Hawea,

The ancient road bridge over the Waitaki River at Kurow is a 'mover and shaker', though not in the usual sense of the phrase! Surely a candidate for replacement in the near future.

Braided Rivers and Hydro Lakes of Canterbury

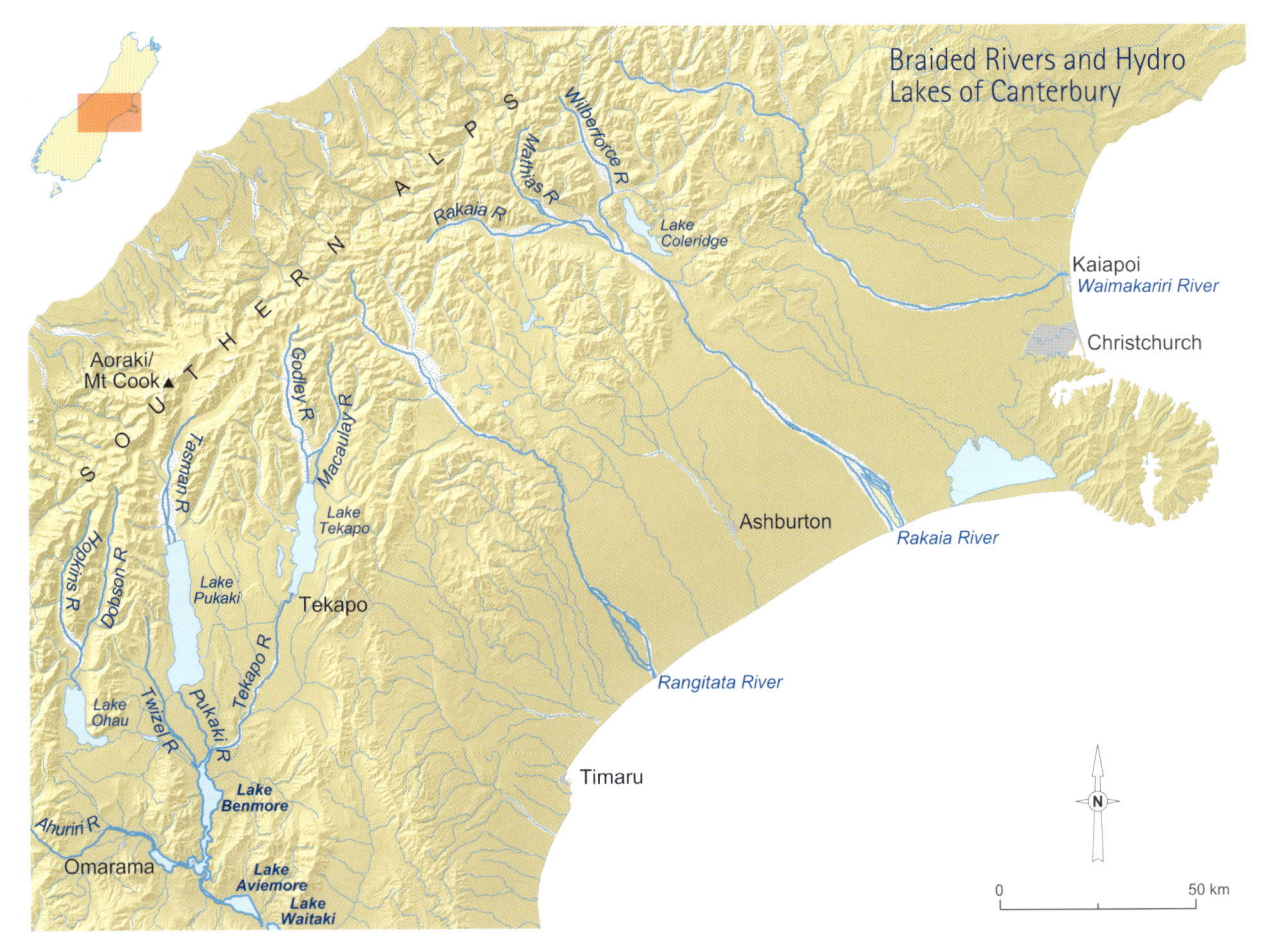

Evening light over the wetlands in the Ahuriri River Valley.

the Ahuriri and its surrounding landscape epitomise everything that is spectacular about this part of the country. It has long been a favoured destination for anglers, trampers and hunters, who, with the consent of the landowner, have ranged widely in its remote regions. In winter, snow patterns the valley floor and the surrounding mountains are white. In summer, tussock grasslands turn the colour of ripe wheat and beech forest drapes the slopes in rich green. Mountain peaks, capped with snow all year round, provide a stunning backdrop.

The 30-kilometre-long valley, traversed by the glass-clear braids of the Ahuriri, is a rare, alpine-valley wetland. With cattle now excluded from grazing the flat lands and the subsequent protection of the habitat for wading birds, the way should be clear for an increase in the population of endangered species such as the black stilt, black-fronted tern, wrybill and banded dotterel. Other species such as the pied stilt, oystercatcher, marsh crake, scaup, black-backed and black-billed gull should also benefit from the enhanced environment.

RIVERS OF THE EAST COAST 103

Hydroelectric power

Waitaki is a name that has become synonymous with hydroelectric power. The first power station in the region was the Lake Coleridge scheme, started in 1911 with the aim of providing the growing city of Christchurch with a regular supply of electricity. Commissioned in 1914, it was the government's first major stake in the power industry and its construction was a notable achievement at that time, built as it was on the loose gravel of the Rakaia River. Knowledge gained on this project helped with the later construction of the Waitaki power station.

The Waitaki Scheme consists of a total of eight power stations, three on the lower Waitaki and five comprising the Upper Waitaki Power Scheme. The first of the dams, the Waitaki Dam, was started in 1928 but was delayed by massive flooding and did not generate power until 1935. It was the last in New Zealand to be built by pick and shovel and wheelbarrow and has an interesting social history. Constructed during the 1930s Depression, it actually trialled the first social security system in the world. The doctor in Kurow at the time, Dr D G McMillan, agreed to provide free medical treatment to the men and their families if they paid a small weekly sum into a common fund. Later McMillan became a cabinet minister and, with Arnold Nordmeyer, also in cabinet (but formerly Kurow's Presbyterian Minister), helped instigate a similar scheme nationwide.

Benmore Dam, also on the lower Waitaki and constructed in 1964, is one of the biggest earth dams in the Southern Hemisphere and its lake is the biggest artificial lake in the country. It is the site of the High Voltage Direct Current link for electricity transfer between the South and North Islands. Nearby Aviemore Dam was commissioned in 1968.

The Upper Waitaki Power Scheme had its origin as early as 1904 when the Hay Report highlighted the potential of the Mackenzie Basin. The first stage started in 1938 with the digging of an outlet tunnel from the south end of Lake Tekapo. Interrupted by World War II, Tekapo A power station was not finished until 1950.

With the completion of Benmore and Aviemore, work started again on the Upper Waitaki Scheme at the end of the 1960s. The first step was the creation of the custom-built town of Twizel. By 1985 the scheme had expanded to include the building of four more power stations and six canals. It had created Lake

The large earth dam and pre-stressed concrete penstocks of the Benmore hydroelectric dam are unique features of one of the major engineering works on the Waitaki River, viewed here through a penstock section on display.

The spillway of the Benmore Dam.

Ruataniwha, raised the level of Lake Pukaki – the major water storage for the Waitaki Power Scheme – and integrated Lake Ohau into the system. The Tekapo B station was built on dry land in the middle of Lake Pukaki. When the lake level was raised it effectively became an island with two thirds of the station below water. The Ohau A station, completed in 1979, is situated on the Pukaki-Ohau canal and connects the water from Lake Ohau with water from both Pukaki and Tekapo. Water from Lake Tekapo passes through eight power stations, before reaching the sea.

As always, hydroelectricity schemes create spin-offs for the nearby communities, not all of them detrimental. The town of Twizel had been intended as a temporary settlement, to exist only for the life of the construction of the power projects. Residents thought otherwise and it is now a small thriving town, acting as a tourist centre and also serving DOC as a regional office. Their black stilt management programme is situated in the town.

An international rowing course has been set up on Lake Ruataniwha. Every second year the New Zealand rowing championships are held there while the Maadi Cup for secondary school rowers takes place in the intervening years.

More recently, Meridian Energy has been involved in a project designed to enable elvers, or young eels, to travel up the Waitaki and for adult long-finned eels to travel downstream and out to the ocean to their ancestral breeding grounds.

Hydro lakes

Travelling from north to south, the first of the turquoise-coloured glacial lakes is Tekapo, originally called Takapo. A seasonal hunting ground for early Maori, they established a pa on Ram Island or Motuariki in the lake. They travelled to Tekapo to harvest lamprey eels from the lake and weka, which were plentiful in the area.

It is a large, deep, glacial lake with a steep shoreline, except at the northern end where the lake begins as a wide delta of gravel and mud flats. Drawdown by Tekapo Power Station in winter exposes shoreline bays and deltas that are particularly important for waterfowl breeding (black stilt, banded dotterel, grey teal and shoveler) and feeding. The lake also harbours brown and rainbow trout.

On a clear day, further south, State Highway 8, which skirts Lake Pukaki, offers fantastic views of Mount Cook across its shimmering expanse of bright water. A large, deep, glacial moraine lake, it is part of the Upper Waitaki Power Scheme and the water level is controlled by a dam at the south end of the lake.

Unlike Pukaki and Tekapo, Lake Ohau is not predominantly glacier-fed, although there are small glaciers in the heads of the main valleys and the water is clear and cold. It occupies the lower end of a glaciated valley, confined by a moraine 16,000 to 18,000 years old.

The name Ohau occurs frequently among Maori place names, and the meaning usually accepted is the literal one, 'windy place'. Long-finned eels live in the lake and it provides excellent fishing for brown and rainbow trout.

In summer, the blue of the lake makes a sharp contrast against the stark hills that surround it. Snow arrives late in autumn when the many larch trees near the water turn into golden torches lighting up the dark foliage of the neighbouring pines.

A small skifield lies up a tortuous road above the lake, but Ohau is still off the main tourist track and, with its tiny village, retains the back-country grandeur that Wanaka and Queenstown have long since lost and which is fast changing at Tekapo.

The three lakes provide feeding, roosting and breeding habitats for open water divers, deep and shallow water waders, wrybill and banded dotterel. Gulls, terns, black stilts and the southern crested grebe are also frequent visitors.

Winter magic along the Tekapo hydroelectric canal.

ABOVE: Few people do not marvel at the beauty of Lake Pukaki. Glacial rock flour suspended in its waters gives it a unique colour.

OPPOSITE: Larches in autumn and early snow on the hills reflect in tiny Lake Middleton, adjacent to Lake Ohau.

RIVERS OF THE EAST COAST

The braided Godley River, feeding into Lake Tekapo — one of the memorable sights seen when flying from the tiny airport near the lake.

A crisp winter morning at the tiny fisherman's village on the shores of Lake Alexandrina, near Lake Tekapo.

OPPOSITE: Mount Cook in all its glory looms over the Hooker River on the walk up to the terminus of the Hooker Glacier.

A LAND OF WATER

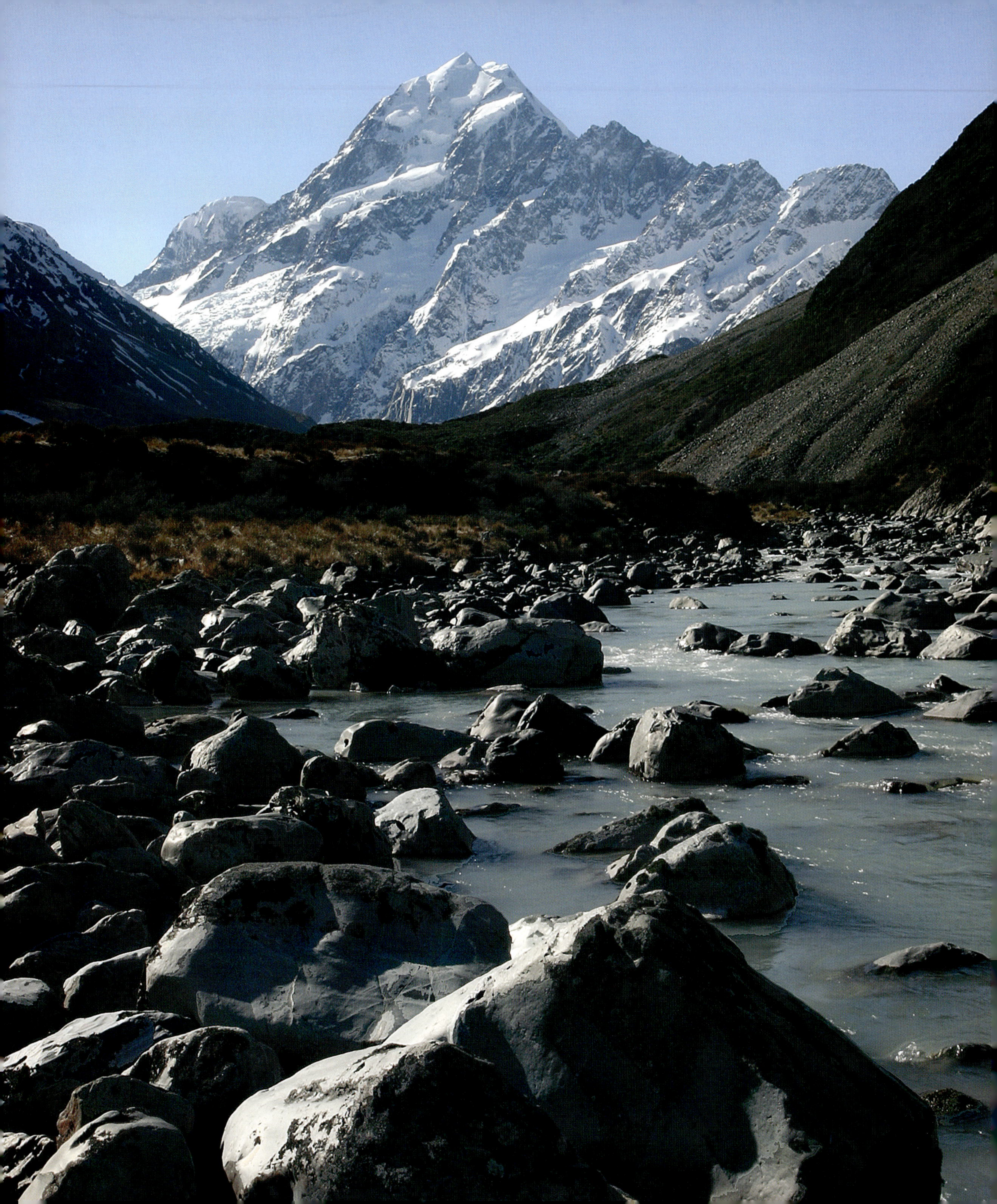

BORN OF GLACIERS

OTAGO'S LEGENDARY LAKES AND MOUNTAIN RIVERS

A benign snow-fed river for most of the year, the Matukituki forms a backdrop for trampers on the climb to Rob Roy Glacier.

It's hard to imagine the changing faces of New Zealand through millennia. Shaped by radical climate fluctuations, volcanic upheaval and the action of wind, water and fire, the land and its contours have altered unrecognisably. Add in changes made by humans in more recent times, though, and the reality of a dynamic landscape becomes easier to understand – and often more difficult to accept. The creation of Lake Dunstan is a ready example.

In contrast, the disappearance of Lake Manuherikia happened before humans were around to notice. Few people are even aware it existed. Formed perhaps 15 million years ago, it is surmised that the lake stretched from Ranfurly in the east of Otago to the Remarkables in the west. Before that time, sea had covered much of New Zealand. When it receded, water was trapped inland and a swampy plain remained in the Maniototo and its environs. Judging from fossil findings, the climate was subtropical, and a fragment of jawbone from a freshwater crocodile discovered in 1989 near St Bathans suggests a type of everglades environment very different from the parched region that is Central Otago today.

The relatively new lakes of Wakatipu, Wanaka, Hayes and Hawea owe their existence to a very different process. Until the last ice age ended about 12,000 years ago, glaciers covered much of the Southern Alps. Constantly advancing and retreating, they carved out trenches in which the lakes now reside and their gouging action was responsible for the extraordinary depth of both Lake Wanaka and Lake Wakatipu.

The glaciers also left behind them heaps of terminal moraine, still evident at Kingston at the foot of Lake Wakatipu, where an ancient river once drained the lake and joined the Mataura River system. Now the Kawarau drains Wakatipu from Frankton and is a tributary of the Clutha. In the past, the Shotover River emptied directly into the lake. Now it is a tributary of the Kawarau and thus of the Clutha also.

One Maori legend ascribes the formation of Lake Wakatipu to Te Rakaihautu and his busy digging stick.

A LAND OF WATER

Another claims that Wakatipu was formed when a giant ogre burned and sank, causing an enormous hollow that later filled with water. Some suggest that the correct name should be 'Waka-tipua', the separate parts meaning 'trough' and 'goblin'. The trough is the lake, and the heart of the goblin lying at the bottom still beats, causing the lake to rise and fall in a regular rhythm, a phenomenon known to scientists as a seiche. They ascribe this seemingly tidal movement to the relationship between the reach of the lake, its depth and the reactions of wind on its surface.

Early journeys to the lakes

Myths mingle with fact, but remains of 'umu ti' – 13th-century cabbage-tree ovens probably used to extract a sugary substance from the cabbage trees – have been uncovered near the mouth of the Dart River at the head of Lake Wakatipu.

Later, moa hunters took up seasonal residence on the shores of the lakes, sometimes staying for months at a time. Vestiges of campsites have been discovered on the banks of both the Dart and Rees rivers at the head of Lake Wakatipu and on Pigeon Island, opposite the Greenstone River Valley. A large Maori pa once stood on the peninsula where the Queenstown gardens now stand.

Seasonal journeying continued into the 19th century, and it seems that Maori knew a variety of routes from the lakes that took them over passes to both North Otago and the West Coast. Maori travelled from Wakatipu over the saddle between the Arrow River and the Motutapu and from there down to the shores of Lake Wanaka where kaika or small settlements existed. At the head of Lake Wanaka, they also followed the Blue River from Makarora over Maori Saddle and so to the Okuru River on the West Coast.

Irvine Roxburgh, in his book *Wanaka Story*, relates how marauding northern Maori allied to Te Rauparaha found their way into Central Otago from the West Coast in 1836. They killed or took captive most of the Maori living at the time around both Lake Wanaka and Lake Hawea, but when the raiding party continued south to the coast they were defeated and killed. When Europeans were pushing through to the interior of the country a few years later they found it empty, though remnants of Maori cultivation on the western shore of Lake Wanaka near Roy's Bay still existed when the Normans, the first Pakeha family to settle in Wanaka, arrived there in 1860.

If the Maori came to the area seeking food and tool-making materials, the Europeans came in search of land on which to graze their sheep. Tales circulated of the large empty land, already burnt bare of forest, with potential for grazing. In the late 1850s, sheepmen arrived, coming over the Lindis Pass from the north, to take up huge tracts of land in the vicinity of lakes Hawea and Wanaka. The names of several are remembered in landmarks of the area, such as Mount Roy, the Wilkin River, Mount Burke and Mount Maude.

Meanwhile, on the Wakatipu side a similar story was unfolding. Following in the wake of the Maori, in 1859 the intrepid Donald Hay built himself a koradi raft (made from dried flax flower stalks) hoisted a blanket as a sail and set off for several days exploring the waters of Lake Wakatipu. A year later, William Rees, he who gave his name to one of the rivers and valleys at the head of Lake Wakatipu, arrived with Nicholas von Tunzelmann, each subsequently laying claim to land for a large sheep run.

Early days in Queenstown

Had gold not been discovered at almost the same time the graziers arrived, the history of Queenstown and the surrounding area may well have developed differently. It would certainly have happened more slowly.

A favourite river for jet-boaters, the Dart River sweeps under the road bridge on the road between Glenorchy and Kinloch at the head of Lake Wakatipu.

BORN OF GLACIERS

BELOW: Spring comes to Cecil Peak and Mount Nicholas, towering over Lake Wakatipu.

BOTTOM: Ruby Island in Lake Wanaka, with autumn colours blazing.

By 1862, miners came in their droves to the Wakatipu area, lured by the promise of gold. During the next few years rich finds were reported on the Shotover and the Arrow rivers, in Skippers Canyon, and on the Moonlight and Moke creeks. On these two latter streams alone, at the peak of the gold rush, more than 3000 men were reputed to be living along their banks, mostly in tents. Winter floods were devastating, and in 1863 many of the miners camped on the beaches of the Arrow and the Shotover were drowned.

The government of the day revoked the grazing licence they had awarded Rees for the land where Queenstown now stands – eventually paying him compensation – and declared the area a goldfield. Rees turned his business acumen to providing for the needs of hordes of hungry men. In 1862 he owned a whaleboat, the sole means of transport on the lake, which plied the route from Kingston at the south end to the burgeoning new town. He ferried provisions to the miners and hauled out the gold they had found. The town grew fast. Hotels sprang up on most street corners, timber milling and flour milling industries were developed, and in 1866 Queenstown became a municipal borough.

By the time the gold was dwindling, people had discovered the beauty of Lake Wakatipu and its surrounding mountains. During the 1870s and 1880s, tourism developed rapidly. Road access was still difficult, but once the railway extended to Kingston numerous steamers plied backwards and forwards to Queenstown. They brought provisions to the town, and visitors from as far away as London and New York to stay at Eichardts Hotel, a Queenstown icon that still stands on the lakefront. In 1878, the basement of the hotel flooded, furniture floated around the lounge on the ground floor and the water was up to two metres deep on some streets. Disastrous floods in the late 1990s saw similar levels of water swirling through the town.

In spite of Queenstown's popularity as a visitor centre, landowners and miners living in the rural areas surrounding it remained isolated for decades. High-country sheep stations in the remote valleys radiating out from the lake had no road access. The people who lived and worked on them were totally reliant on the lake steamers, or sometimes smaller private boats, for all their supplies, for transporting wool and stock and for bringing the doctor in times of emergency.

Few visitors to Queenstown today realise that the picturesque old steamer that takes tourists across to Walter Peak was once a lifeline on the lake. Commissioned in 1909, the *Earnslaw* was a response to tourist demand for a more comfortable service than that supplied at the time by numerous steamers on the lake. Making its maiden journey in 1912, it operated for almost 25 years as a cargo ship while also offering first- and second-class facilities for passengers, including dining room, lounge and bar. But in 1936 the road from Kingston to Queenstown was completed and this dealt a heavy blow to the steamer trade. In

A LAND OF WATER

1952 the *Ben Lomond*, a feature of the lake for 80 years, was withdrawn from service, and only the *Earnslaw* remained to provide a thrice-weekly service carrying goods, dogs, shepherds, tourists, trampers, royalty on one occasion, motorcars, and even buses going to Glenorchy to provide sightseeing tours from there to Paradise in the Dart River Valley.

I remember picnic trips on the *Earnslaw* during Christmas holidays in the 1950s. The boat stopped at various sheep stations along the way; wool might be loaded on or a gang of shearers be dropped off and there was always time for a swim at Elfin Bay at the mouth of the Greenstone River. New Year's Eve, in the same period, and the prosaic became romantic when the *Earnslaw* departed on moonlight cruises and softly lit figures danced onboard with the shadowy mountains looming in the background.

Shortly after the road to Glenorchy opened in 1968, the *Earnslaw* retired from a distinguished working career to continue as a venerable tourist attraction.

The lakes today

While Queenstown developed by degrees into a resort town with an international reputation, Wanaka, (known as Pembroke until 1940), on the other side of the Crown Range, remained for many years a sleepy village beside the lake, serving a far-flung rural community of mainly sheep farmers. Hawea was even smaller. Summer would see an influx of visitors, usually residents from Dunedin or Christchurch spending their summer holidays in 'Central'. With the development of nearby skifields in the 1970s, the character of the two towns started to change.

Now, in the 21st century, the lakes are the focus of an area that attracts visitors from all over the world. They come to enjoy the magnificent scenery, to participate in winter sports or to take part in some of the most adventurous holiday activities that exist. The first-ever commercial bungy jump was set up by Henry van Asch and A J Hackett in 1988 at a bridge over the Kawarau River near Queenstown. Jet-boating attracts thrill-seeking amateurs as well as highly experienced racing-boat drivers, and the biennial War Birds Over Wanaka event brings anyone to the area who is in love with flying. With world-class golf courses, superb fishing opportunities and numerous intriguing vineyards, it is not surprising that real estate anywhere in the lake district attracts high prices.

The gold generated now from Queenstown and its neighbours is of a different colour from that extracted in a frenzy of activity in the pioneering days of New Zealand.

Mountain rivers

These lakes all drain eventually into the much-modified Clutha River, but the rivers that feed the lakes remain primitive mountain waterways. Snow-fed, fast-flowing, punctuated by boulders and prone to rapid rises, their valleys provide tramping routes into Mount Aspiring National Park, and their waters provide sport for anglers and thrills – sometimes unpredictable – for jet-boaters. Well I remember an expedition into the Wilkin one fine May day. Six day-trippers, a hamper filled with picnic delicacies and a driver whose attention wavered for a vital five seconds. Result: one boat firmly marooned on a gravel bank, three hours of daylight remaining and a three-hour walk to Makarora and assistance. But luck was a lady that day. Fifteen minutes of fast hiking along the riverbank and suddenly there was a muted roar in the distance and several jet boats appeared around a bend in the river with the manpower to drag our boat free. The sun was setting as we finally arrived at our destination, pulled out the lunch hampers and built a driftwood fire to dry out the wet clothes.

Paragliding on Lake Wakatipu – one of the many activities in and around Queenstown, the 'Adventure Capital of the World'.

The busy lakefront at Queenstown.

ABOVE: A magic winter morning at Glendu Bay, Lake Wanaka.

RIGHT TOP: Lake Johnson nestles in the rocky hills on the outskirts of Queenstown.

RIGHT BOTTOM: Jet boats give easy access to the shallow braided rivers of the South Island such as the Dart at the head of Lake Wakatipu.

A LAND OF WATER

BORN OF GLACIERS

An extensive shingle fan in Lake Hawea, built up by the Dingle Burn, is a footnote in the story of our ever-changing country.

RIGHT: Queenstown and Lake Wakatipu from the top of the Skyline cable way.

OPPOSITE RIGHT: A sparkling tributary of the Hunter River at the head of Lake Hawea.

116 A LAND OF WATER

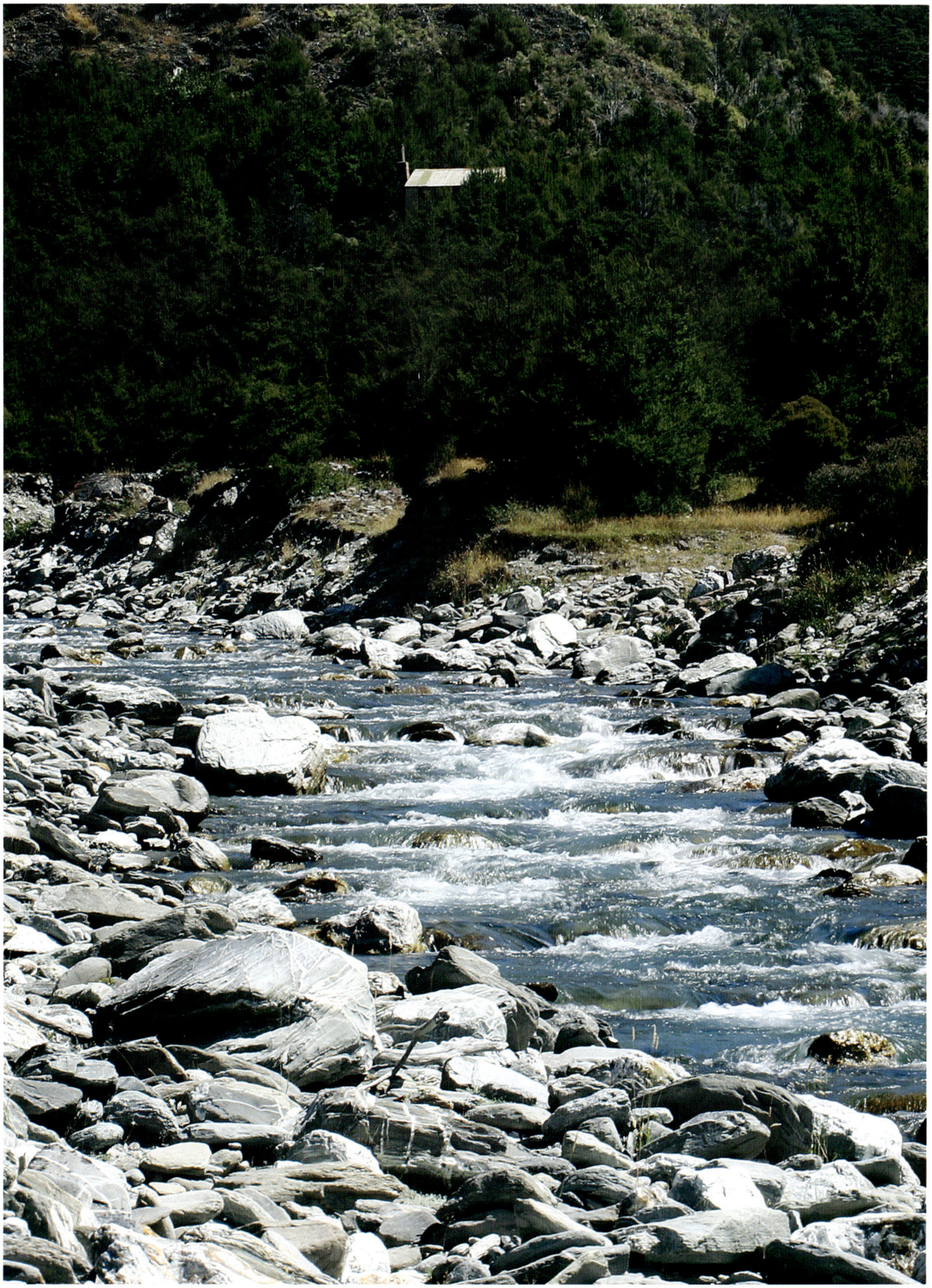

TOP: Matukituki River after an early dusting of snow.

ABOVE: A female paradise shelduck.

RIGHT: This swing bridge gives access to a lodge in the Matukituki Valley, owned by Otago Boys' High School.

118   A LAND OF WATER

ABOVE: The Kawarau River swings past the foot of the Remarkables on its way to meet the Shotover River, before plunging through the Kawarau Gorge.

LEFT: Invasive lupins make a splash of colour in spring in the upper reaches of the Cardrona River, near the Crown Range Road.

THE MIGHTY CLUTHA

BELOW: The mighty Clutha River is still cutting into the flood plain to create distinctive river terraces.

BOTTOM: Anonymous constructions beside the Makarora River at the Blue Pools, reached from the Haast Road beyond Lake Wanaka.

I don't remember when the Clutha River was not part of my life. It featured in bedtime stories my father told me. It signalled our arrival in Central Otago during tedious holiday journeys from Dunedin to Alexandra in the years before I started school. Later, on trips to Queenstown, I marvelled over its hectic course through the Cromwell Gorge, and later still, I mourned its taming when the Clyde Dam was finally completed.

Now we live close to its source, we bike alongside it and the first glimpse of its bright ribbon of colour when we come to Wanaka from the north tells us we are home.

I didn't always know it as the Clutha. When I was a small child, that was the name of the turquoise river that poured down from Lake Wanaka. At Cromwell it met the darker Kawarau and the two-coloured currents tumbled together to become the Molyneux.

The early whalers and settlers of South Otago called the river and their district the Molyneux, though the name Clutha, coming from the Scots Gaelic word Cluaidh, meaning the River Clyde in Glasgow, was apparently first suggested in 1846 by Scottish arrivals eager to enshrine memories of home. At some stage while I was growing up, I came to know the longer river as the Clutha.

Walk along the track above its jade-coloured waters as it leaves Lake Wanaka, trace its path from the bony hills above Alexandra, cross the bridge at Beaumont where you can peer into its translucent depths and the memory of the Clutha will stay with you forever. Tamed it might be, but it still retains the power of enchantment.

The river's course

Second only to the Waikato in length, the Clutha derives from the most complex catchment area in the country and, at more than 20,000 square kilometres, also the largest. In the west it is fed from high rainfall areas of the Southern Alps where bush blankets the lower slopes of the mountains; closer to the coast, innumerable streams from rolling green hills feed it. But for much of its course, the largest river in New Zealand by volume sweeps through the driest landscape in the country, a peacock streak in startling contrast to the parched, tawny country around it.

The Clutha flows out of the southern end of Lake Wanaka, but it starts life close to Haast Pass as an insignificant stream that becomes the Makarora River, emptying into the northern end of Lake Wanaka. Through the Upper Clutha basin, the river is joined by the Cardrona, the Hawea and the Lindis rivers, while its largest tributary, the Kawarau, meets the Clutha at Cromwell. By the time it reaches Balclutha in South Otago on its 320-kilometre journey to the sea, the river has slowed and widened, its stature increased yet again with the inflow of the lower tributaries, including the Manuherikia, the Pomahaka, the Tuapeka and the Waitahuna. Beyond the distinctive Balclutha Bridge it bifurcates and straddles the land to form the Inch Clutha Delta, a flat, fertile island of

A LAND OF WATER

dairy farms crisscrossed by country roads. Repeated floods have devastated the delta in the past, but the 7000 hectares of land are now largely protected from serious damage by a control scheme initiated in 1960, after prolonged flooding in 1957. Flood banks, spillways, tide-gates and drainage pumping stations all play their part in taming the river.

The mouth of the Clutha is elusive when we go looking for it. At the end of Centre Road we eventually see the sign: Mouth of the Clutha. We park and cross a small bridge, following the flood banks along the winding final kilometres of the river. Wind rattles in the flax bushes beside the water. The odour of cow is unmistakable. Seagulls shriek overhead and in the distance there's a glimpse of navy-blue sea. The track curves, and several whitebaiters' shacks come into view. One has slipped, lopsided, into the river. Flax bushes give way to lupins. Water spreads in all directions, and suddenly we're there: the mouth of the mighty Clutha. Or half of it. It wasn't always like this. Prior to the record-setting flood of 1878, the two strands of river shared a common mouth at Port Molyneux, near Kaka Point. We have been following the Matau branch. A couple of kilometres south, the second and larger branch of the Clutha, the Koau, disgorges into the sea.

Such a long river means many things to many people. If you know it near its source, it's a wild river, speaking of the mountains and the special character of Central Otago with its rugged hills and rocky gorges. If you know it near its mouth, it's a gentler river that sweeps in broad curves past fertile paddocks. Anglers, kayakers and jet-boaters love it for the sport it offers; horticulturists and farmers value its waters that nourish their crops. For town-dwellers along its banks it carries memories of flood; to city-dwellers far away it represents possible electricity supplies.

Early journeys

Maori knew the mighty river as Mata-au, meaning 'surface current', and it once marked the boundary line between Ngai Tahu and Ngati Mamoe areas of influence. Its delta area served as a larder; the river itself as a route back to the coast after arduous trips into the interior of the country in search of seasonal food or further afield to find pounamu. There is no doubt that Maori were familiar with Central Otago, even if, as seems evident, they did not set up permanent settlements away from the coast. Moa-hunting sites have been found inland, in a number of areas close to the Clutha. Too turbulent to paddle up on their way to the interior, the river served them well on their return to the coast. They negotiated its strong currents and occasional rapids in rafts known as 'mokihi', made sometimes from 'korari', dried flax flower stalks, and sometimes from raupo, the buoyant stalks of bulrushes.

Nathaniel Chalmers, one of the first white men to penetrate into Central Otago, came to know this kind of raft intimately. In 1852 he hired the services of Maori chief Reko from Tuturau to guide him into the interior from the South Otago coast. Their trip pushed him to the limits of endurance. They followed the Mataura River at first, branched up the Nokomai near present-day Garston, found their way through the rugged Nevis Valley and so to the Kawarau River, apparently crossing it by jumping the natural bridge that is still visible beyond the Roaring Meg Power Station. From there they followed the course of the Clutha River until they reached lakes Wanaka and Hawea. Chalmers' intention had been to continue into Canterbury, but he was so depleted physically and morally that he decided to return to the coast. Reko and his Maori companions constructed a traditional korari raft and the party made a quick, if perilous, journey back to their original destination.

Ferries and bridges

Tales abound of early runholders fording the river on foot and horseback, and later shepherding their flocks

Final destination: the Matau branch of the Clutha meets the Pacific Ocean.

of sheep across shallow fords, if they could be found. Frequently it involved long detours to find a suitable crossing, and the Molyneux posed huge risks with its strong current and high volume. An unwary moment and both men and horses drowned.

As more settlers arrived in Otago, the problem of how to transport people safely across a tumultuous river posed a serious logistics problem. In the very early days of Otago settlement, boatmen set up ferry services at various crossing points, using whaleboats. Travellers on their way south from Dunedin had first to cross the Taieri by private ferry and then the Molyneux at Balclutha. In 1856 the Provincial Government organised ferries in accordance with the Ferryage Ordinance passed that year. Crossings of the Clutha were established at Teviot (Roxburgh), two at Lower Dunstan (Alexandra), one at Upper Dunstan (Clyde) and two at The Junction (Cromwell). Rates were established and the ferry services were put out to tender, including that at Balclutha.

The discovery of gold in the early 1860s meant a stampede of people to the interior of Otago and the number of ferry crossings along the river increased to cope with the traffic. Most of the later models were flat platforms rather like a small bridge, mounted on pontoons and tethered to a permanent wire rope to keep them on course while the current of the river provided the propulsion. As the century progressed, bridges gradually took their place.

At Albert Town, where the Hawea and Cardrona rivers meet the Clutha, two punts operated until 1930, and several generations of schoolchildren in the area enjoyed an unplanned holiday whenever flooding prevented the punt crossing.

The golden era

The tributaries and banks of what was then known as the Molyneux were a rich source of gold. Gabriel Read discovered gold at Tuapeka in 1861, but the greatest rush was in 1862 when Hartley and Reilly made their discovery at the Dunstan, legally proclaimed a goldfield in September of that year. Its boundaries ranged from the junction of the Lowburn and Clutha rivers to the junction of the Manuherikia and the Molyneux and included the land one mile on either side of the Molyneux.

By 1863 it was estimated that between 30,000 and 40,000 diggers worked along the banks of the

A LAND OF WATER

Old miners' huts and sluicing tailings along the Kawarau River give some hint of what life was like for gold prospectors in the gold rush days.

Molyneux and were supplied with all their needs from the two towns of Alexandra and Clyde. In summer, when the river was low, miners camped along the beaches at the river's edge. When it flooded that winter, more than one fortune-seeker was drowned and many lost all their possessions.

The first gold-seekers came with a simple pan or built themselves a 'cradle' to wash the gold-bearing gravel; later efforts, dating from 1870, were centred on sluicing and dredging. Access to water assumed enormous importance and still visible today are some of the water races that were dug by hand to take water from tributary streams of the river to gold claims.

The earliest spoon dredges were soon succeeded by current wheelers, which drove a continuing chain of buckets, using the river current to supply power. Steam dredges first appeared in the early 1880s, and in 1890 the first electric dredge ever built was made in New Zealand and started working on the Upper Shotover River, a tributary of the Kawarau. A gold-dredging boom was born. Claims along the Molyneux, Shotover and Kawarau rivers were taken up like shares in modern software companies. It was a time when young men with a large helping of ingenuity and a willingness to take risks could prosper.

The dredges were unwieldy structures, and manoeuvring them in the treacherous currents, before the Clutha was tamed, required considerable skill. Ropes frequently fouled. Capsizing was a constant risk. Life aboard was noisy, uncomfortable and often dangerous. Alexander Crow McGeorge, part owner with his brothers of the Electric Company, which operated dredges on the Kawarau above Cromwell, describes the following incident in the memoirs he wrote for his family:

One night, about 11 pm, an essential pulley broke. Alex, ready to turn in for the night, remembered having seen an abandoned fly wheel from a dredge in Alexandra. He and his younger brother Joe set off for the town on foot, arrived about 2 am and roused a fellow crew member to take them to the site with his dray. There they hauled the fly wheel up the steep riverbank, humped it onto the dray and drove back to the dredge where Alex modified the wheel to fit. By 8 am the next morning when the new shift, including Alex, was ready to take over, the dredge was ready to work again.

In subsequent years the fantastic returns of both

OPPOSITE: Bridges over the Clutha — at Alexandra, Clyde and Balclutha.

alongside the Clutha, eking out a minimal living for several years, much as the Chinese miners had in an earlier age.

With improved technology, dredging was revived during the 1930s and continued for another 20 years or more, with machines working areas alongside the river that had not previously been accessible.

In the 1980s, when the Clyde Dam was under construction, dredging was once again carried out on some of the beaches and river terraces that were due to be drowned when the dam was filled.

Modern developments

Fortunes have been won and lost along the banks of the Clutha. When the easy gold ran out, many miners departed to seek their luck elsewhere. Others stayed to earn a living – if not a fortune – by establishing small farms or orchards. Initially, apricots and cherries were favoured fruit; more recently, apples have dominated, and grapes are the latest crop to tempt investors' gold.

Access to water has always been a critical factor in the development of the dry Central Otago region. It was the cause of many disputes during the mining era. When water was no longer needed for mining, many of the original rights were transferred to those needing water for irrigation purposes. Today, the competing rights of hydroelectricity producers, recreational fishers, and farmers and orchardists along the course of the Clutha are still the cause of legal battles.

Man-made changes to the river itself have also caused much strife. When the Roxburgh hydro dam was completed in 1956, most people viewed its presence as a sign of progress. But by the mid-1960s attitudes had diversified as people became more aware of the need to protect natural landscapes. When plans surfaced to build a second dam at Clyde, involving the drowning of the old Cromwell township, a long and bitter battle ensued. In spite of local and national protest, plus construction difficulties, the dam was eventually completed and Lake Dunstan was filled in 1993. The result was massive changes to the landscape. Along with the destruction of the old part of Cromwell, long-established farms and orchards were drowned and the rugged character of the Cromwell Gorge was altered forever. Where a turbulent river had raged, a tranquil lake was formed.

Regrets still linger among those who knew Cromwell and the river before the dam was built, but

TOP: Water anything and farmers can work miracles: a modern pivot irrigator, nearly a kilometre in length, brings water to formerly dry river terrace paddocks near Tarras.

ABOVE: Power for the people: the Clyde Dam at full generation.

the Electric Company, which operated dredges on the Kawarau above Clyde, and the Hartley and Reilly dredge on the Molyneux helped trigger a scramble to form gold-dredging companies. In the first years of the 20th century, some records claim that more than 150 dredges were operating on the Molyneux, Kawarau and Manuherikia rivers.

Such activity was relatively short-lived. A gradual decline in gold-seeking over the following decades was briefly revived during the Great Depression when unemployed men made their homes in primitive huts

A LAND OF WATER

Early winter reflections: Cromwell mirrored in the waters of Lake Dunstan.

the filling of Lake Dunstan and the establishment of new facilities in Cromwell, such as a campus of Otago Polytech, have coincided with a burst of tourist development in Central Otago. Irrigation in some areas has been facilitated by agreements built into the operation of the Clyde Dam and the new lake has proved to be a welcome addition to the area as a venue for wind-surfing and trout-fishing. It has also played an important role in providing irrigation for the vineyards that have been established since 1990 in the Bannockburn region and along stretches of the Clutha between Wanaka and Cromwell.

Water, however, can be a mixed blessing. Flood waters have inundated areas in and around Alexandra six times in the last 25 years. In the 1990s a succession of floods swept down the Clutha bringing with them widespread damage and heartache, along with new controversies. The flood of 1999 was the second-biggest flood reported on the Clutha, yet the river rose higher in Alexandra than it had during the greater flood of 1878, causing enormous disruption in the community. Local feeling ran high, blaming the build-up of sediment in Lake Roxburgh and the consequent backing up of the Manuherikia River where it meets the Clutha.

In 2000, the government, in tandem with Contact Energy Limited, agreed to pay compensation for damage sustained and also to pay for substantial flood protection. Concerns are also mounting over the rapid accumulation of sediment in the Kawarau arm of Lake Dunstan and the subsequent effect on both irrigation and water storage.

TOP: The Kawarau River surges through the gorge at the natural bridge, once used by Maori travelling from the east coast of Otago to the West Coast.

ABOVE: Willows in early leaf along the Clutha, downstream of Luggate.

ABOVE: Wild and remote, the Shotover River courses through Skippers Canyon in the rugged country behind Coronet Peak ski-field.

TAIERI RIVER AND ITS CATCHMENT

ABOVE: The Scroll Plain in the upper reaches of the Taieri River can only be fully appreciated from the air.

BELOW: A veteran of the Taieri River, this old bridge is not far from Dunedin.

If you grew up in Dunedin, 'The Taieri' is one of your compass points. It may mean the river or it may mean the plain that the river drains. It's where you go to catch a plane, catch some sunshine or head south.

The course of the river

The Taieri River rises in the Lammerlaw Range among the golden tussocklands of the newly established Te Papanui Conservation Park, one of only two grassland parks in the country. It contains broad peaty basins and a succession of streams dissect its slopes. It is significant for the Otago region, protecting a large proportion of Dunedin's catchment area.

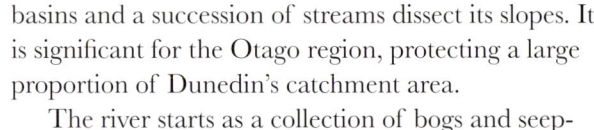

The river starts as a collection of bogs and seepages, gradually gathering into a stream. In all it drains a total of 5650 square kilometres. If it flowed in a straight line, it would be a short run to Taieri Mouth on the coast, about 45 kilometres south of Dunedin. But the terrain decides otherwise, and the Taieri, at nearly 320 kilometres, is the third longest river in the country, meandering along a convoluted course north before making a U-turn around the end of the Rock and Pillar Range in the Maniototo. From there it heads south through Strath Taieri and the deep-cut Taieri River Gorge, collecting fast-flowing tributary streams as it goes. Across the Taieri Plain it's a peaceful river most of the time, flowing quietly to the sea, with branches into Lake Waihola and Lake Waipori.

When Maori roamed this area it was a different world. In the words of David Monro, who explored here in 1844, it was 'an immense grass-tree swamp, through which canals of black sluggish water wind in various directions and interspersed with stagnant lagoons …' Rather different is the green fertile farming area we know today. With a catchment of almost 20 percent of Otago, the river has always wielded considerable influence on the life of the region.

128 A LAND OF WATER

Early days on the Taieri

If you were an early settler in Dunedin, the Taieri hemmed you in. No sooner had you headed south or southeast and your progress was hindered by water. The area south of Dunedin was, for a period, totally dependent on river communication and transport. Coasters from Port Chalmers and Dunedin sailed to Taieri Mouth and from there proceeded up the Taieri River to a jetty at Lower Taieri Ferry, now Henley, where goods were transferred to a punt for onward conveyance. When gold was discovered in the hinterland, men in their hordes headed into the interior, and the Upper Taieri Ferry at Outram became a bottleneck. There were reports of crossings day and night, with wagons strung out for a couple of kilometres waiting to cross. It took several more years, however, before the ferry was replaced by a bridge.

It was even longer before a regular ferry operated at Taieri Mouth, improving access to the south. So treacherous was the bar at the entrance to the river, that local operators who knew the waters were needed to guide boats through the ever-changing channel. Between 1862 and 1864 two pilots and their families lived on the tiny windswept island of Moturata at the mouth of the river. A ferry was finally installed in 1886, allowing goods to come directly south from Dunedin, but not until 1912 was it replaced by a bridge.

Today's river

Then as now, the Taieri River was prone to serious flooding, to which the Waipori River contributes. The Waipori also rises in the Lammerlaws, meeting the Taieri at Henley and spreading through the wetlands around lakes Waihola and Waipori.

Some areas of the Taieri Plain lie permanently below sea-level and rely on pumping systems to remain productive. A disastrous flood in 1980 inundated the whole area, and the airport was out of action for 53 days. More recently, serious concerns have been raised about water volumes in the river and its quality. The upper Taieri is one of the driest areas of New Zealand, which traditionally supported only high-country sheep and beef farming. Today irrigation schemes have turned dry tussock grassland into large dairy farms. A monitoring programme in 2001 and 2002, instituted by the Otago Regional Council, showed that swimming holes in the remote Maniototo, where I used to swim as a child, are now often unsafe; that nutrients, bacteria and suspended solids have increased in some areas; and that low flows, already a problem in drought years, were exacerbated by water being drawn off for irrigation. On the Lower Taieri Plain problems include polluted run-off from farms, septic tanks and urban storm water.

The Taieri Trust, an environmental management initiative, was set up with community support and input from Otago University. Its goal is to improve the health of the Taieri River's waterways, and already other communities beyond the confines of Otago are using it as a model in dealing with similar issues. It is hoped that the varied groups with a vested interest in the benefits from the river will come together with a set of compromises that will restore the river's water quality.

BOTTOM: Gripped by hoar frost, the lonely Poolburn Dam in the upper Taieri area is deserted. In summer fishermen arrive to try their luck.

TAIERI GORGE RAILWAY: Departs daily from Dunedin and travels to Middlemarch through the rugged and spectacular Taieri River Gorge.

OTAGO RAIL TRAIL: Biking and walking along the former railway line from Sutton to Clyde. This trail follows the Taieri in places.

LAKE WAIHOLA: Situated 40 kilometres south of Dunedin. A favourite haunt of yachtsmen, waterskiers, power-boaters and anglers, with whitebaiting (in season). Safe swimming and sheltered picnic areas. Trout fishing.

MANIOTOTO: Ice-skating, curling.

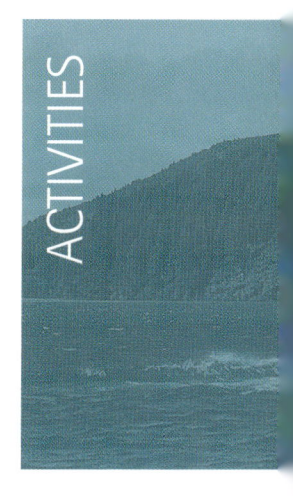

ACTIVITIES

BORN OF GLACIERS

ABOVE: A legacy of enterprising city forefathers, the No. 4 dam on the Waipori River near Dunedin. This small hydroelectric scheme has provided the city with electricity since the early 1900s.

RIGHT: A train ride through the Taieri Gorge is a popular day trip from Dunedin.

Quiet days at Henley on the Taieri Plain.

Rare these days, a curling bonspiel can only take place when a succession of hard frosts freezes the Idaburn Dam near Oturehua in the Maniototo.

BORN OF GLACIERS 131

THE FAR SOUTH

LAKES AND RIVERS OF THE SOUTH

A gathering storm descends on Lake Manapouri.

Water is the defining characteristic of the southwest region of the South Island. It grows the forest. It pours down the mountains to create innumerable rivers and streams. It plummets over cliff faces to form waterfalls and cataracts that are unimaginable if you haven't visited Fiordland during prolonged rainstorms.

It pools in countless lakes that dot the landscape. Some are so small they don't warrant a name on the atlas, others are significant in size as well as importance. Lake Te Anau is the second-biggest lake in the country and Lake Hauroko, at 462 metres, is the deepest. Manapouri, dark, moody, surrounded by dense bush and scattered with myriad, sometimes mysterious, islands is considered by many to be the most beautiful in New Zealand. It is also the lake that has aroused the greatest controversy. Lake Widgeon bears a name that most people think is a case of mistaken spelling and Sphinx Lake sounds as if it should be in Egypt. How many people have laid eyes on Lake St Patrick in the Townley Mountains? How many New Zealanders could place Lake Poteriteri on a map – even though it's larger than Lake Rotoiti in the North Island?

All these lakes contribute to the wilderness nature of this part of New Zealand and are highly valued for their aesthetic qualities. Governments have learned to tamper with them at their peril.

Why so many lakes so far south? The answer, as with many lakes in the heart of the Southern Alps, lies in the effects of glaciation. When the ice retreated from the massive granite rocks of Fiordland and the metamorphosed schists of South Westland, it left behind depressions where water accumulated. Often the water was all but blocked by sills of rock. Little pockets of water, with only a thin stream as their outlet, are the result. Lake Quill on the Milford Track, the source of Sutherland Falls, is perhaps the best-known New Zealand lake of this type. Most are secreted in remote mountain fastnesses, visible only from the air or by hardy trampers and hunters who fight their way through bush in search of true wilderness.

The few well-known lakes now attract tourists from around the world, but when Europeans first started exploring New Zealand, it was several decades before they penetrated to the interior of this part of the country and came across the lakes for the first time.

Of course, Maori had long visited them. Legend has it that when one of the original canoes, the *Takitimu*, under the command of the chief Tamatea, was wrecked in Te Waewae Bay at the bottom of the South Island, the survivors headed inland following the Waiau River. Eventually they arrived at Lake Te Anau where they camped temporarily. In later times a pa, most likely used during seasonal visits, was established on the shores of the lake where the Upukerora River enters the lake, not far from the present-day town of Te Anau. Groups of coastal dwellers would visit to hunt eels and birds. Richard Henry – in 1883 the first settler on the future town site of Te Anau, and later appointed to the Resolution Island reservation as the world's first ranger for endangered species – examined ovens at the old pa and found charred moa bones. Te Anau also figures in Ngai Tahu histories as the site of the last battle between the early rival tribes of Ngati Mamoe and Ngai Tahu.

It was two Maori guides who accompanied explorers W H Stephen and C J Nairn inland to Lake Te Anau and on to Manapouri in 1852 – or as Nairn named it at the time, Lake Moturau, meaning lake of many islands.

Pioneers followed on their heels, keen to appropriate land for farming. In the early 1860s surveyors arrived to survey both Lake Te Anau and Lake Manapouri in their entirety – no small task given the irregular shape of Lake Te Anau and the number of islands in Lake Manapouri. In the course of their survey they also ventured further south to put lakes Monowai and Hauroko on the map.

One hundred years later, Lake Hauroko became a news item with the discovery of a Maori burial cave on Mary Island, complete with a skeleton still wrapped in a cloak. Carefully respected by those who investigated the find, the cave is protected, preserving the tapu of a place holy to local tribes. Scientific estimates date the skeleton back to approximately the time of Tasman, making it one of the oldest burial caves in the country.

Kayakers seeking a true wilderness experience come to Lake Hauroko to try their craft and skills on the Wairaurahiri River, which drains the lake, falling 185 metres in its rush to the Tasman Sea west of Te Waewae Bay. It's a lonely river. The banks are bush-covered and littered with logs, moss-covered trees overhang the current and the water is a dark, peaty brew.

Trampers also visit this area. The Dusky Track begins at the western end of Lake Hauroko, linking it with Lake Manapouri, and the Hump Track links Te Waewae Bay and Lake Hauroko passing above the bush line. On a fine day there are spectacular views over the south coast and inland lakes.

Curving across a valley like a giant boomerang, Lake Monowai is a picturesque lake well known to Southland anglers, hunters, campers and trampers. It is also the site of one of the most scenic hydro power stations in New Zealand. Built to provide Southland with electricity in the days before the national grid existed, it is still going strong.

Lake Te Anau

Camp beside Lake Te Anau in autumn and it's likely you will awaken to a world shrouded in mist. As the sun warms the land, bush-covered mountains reveal themselves across the dark water. You may hear the occasional bellbird fluting through the silence but little else disturbs the tranquillity of this lake, which has so far escaped the developers' bulldozers.

Imagine a hand with a skinny, foreshortened palm, turned on its side, with four tapering fingers stretching into the rugged wilderness of Fiordland. This is the shape of the lake. The thumb of the hand embraces the

A calm evening at Lake Te Anau.

township and the southernmost bay where the Waiau River starts on its journey to Lake Manapouri. With four distinct 'arms' or fiords, it is a paradise for boaties.

Largest of the southern lakes, it was formed between two and 10 million years ago, is up to 417 metres deep in places and owes its lanky, irregular shape to glacial origins. Given its mountainous situation, numerous streams and rivers feed the lake, but the Eglinton River, flowing into the lake north of Te Anau Downs, is its largest contributor. At the northernmost point the Clinton River enters the lake, and this is where walkers set out to tramp the Milford Track.

Much early exploration of Fiordland centred on Te Anau. In 1889, Richard Henry, with Robert Murrell of Manapouri, paddled up the middle arm of the lake, portaged their canvas canoe over the narrow neck that separates Lake Te Anau from Lake Hankinson, continued across Lake Thomson and from there cut a track through dense bush over the subsequently named Henry Pass and so found a route to George Sound.

Two years earlier, Quintin Mackinnon, with a companion, found his way from the South West Arm of the Middle Fiord of Lake Te Anau across to Caswell Sound. In 1888 he discovered and crossed the Mackinnon Pass, the first practicable land route to Milford Sound and already recognised for its unique beauty. During the next several years Mackinnon spent much of his time developing the Milford Track and guiding visitors, including the first woman to walk the track. Sadly, Mackinnon was drowned in 1892, and after an interim when private guides worked the track, the government took over its administration. As Fiordland became more accessible, tourists started coming to the area, and in 1890 the first hotel was built near Lake Te Anau, thus setting in motion the beginnings of the town.

Today, the bulk of the tourist traffic to Milford Sound is inexplicably centred in Queenstown, and Te Anau remains a small town, servicing the needs of farmers in the surrounding district and those prepared to take time to explore the hinterland.

Visitors to Lake Te Anau should not leave before taking a trip to the Te Ana-Au Caves on the side of the lake opposite the town. Although it is thought that early Maori knew about them, and some historians believe that the name for the lake may have been derived from the name for the caves, they were only rediscovered by a local explorer in 1948 – not surprising when you visit them. The entrance is small, close to lake level and surrounded by bush. Hollowed out over millennia by the underground river that rushes through them, the caverns set the scene for a mysterious and awe-inspiring ride by punt into the heart of a cave system that features whirlpools and magical waterfalls, with the eerie blue light of glow-worms the only illumination in some places.

Another fascinating discovery was made in 1948. In that year, takahe were sighted and photographed by Dr G Orbell of Invercargill. He was hunting in a remote valley in the Murchison Mountains to the west of the lake when he came across a bird fossicking in the alpine tussock grasslands of the area. There had been only four recorded sightings of takahe in the previous century – one of which is preserved in the Otago Museum – and it was feared that they had followed the moa to extinction. Following Orbell's dramatic find, several pairs were discovered and a special area within Fiordland National Park was set aside for the takahe's conservation. Now small colonies of takahe are slowly increasing in the security of a few predator-free islands around the country. The Department of Conservation also runs a captive breeding and rearing programme at Burwood Bush Breeding Centre on the shores of Lake Te Anau, where several pairs are held to form a small breeding group. Chicks are reared with minimal human contact; 'handling' is through puppets and models.

Lake Manapouri is a lake of many bays and arms and it contains more than 34 islands which can be explored by boaties.

A LAND OF WATER

'Excess' eggs from wild nests are also managed at the unit to produce birds suitable for freeing back into the wild population in the Murchison Mountains.

Across the road from the wildlife centre is Ivon Wilson Park – 35 hectares of native and exotic trees with paths, picnic areas and barbecue sites. Picturesque Lake Henry is stocked with rainbow and brown trout as a children's fishery. The park is on the way to the control gates that regulate water flows between lakes Te Anau and Manapouri for the West Arm hydroelectric power station and also mark the beginning of the Kepler Track, a three- or four- day walk. The control gates are an easy bicycle ride from the town, and with luck you may see a takahe or two fossicking in their enclosure.

Lake Manapouri

Like its close neighbour Te Anau, Manapouri is a lake of many bays and arms with its furthermost point extending far into Fiordland. Where the Spey River flows out of West Arm, it's a day's walk over the Wilmot Pass and down to the head of Doubtful Sound. In the days before the road was built to facilitate the building of the power scheme we hiked through there and slept under the stars, fighting sandflies by day and mosquitoes by night.

Manapouri is a brooding lake, which echoes the character of its surroundings of misty, bush-covered mountains. Early Maori called it Roto-ua, the rainy lake, which may seem particularly apt to some. Later it became known as Moturau or many islands, also very descriptive as there are more than 34 islands scattered across the lake. The present name is a variation of Manawapouri, which is usually translated as 'Lake of the Sorrowing Heart'.

The development of the town as a small tourist centre is synonymous with the name of Murrell. The first Robert Murrell arrived in New Zealand in the 1860s and settled nearby and his descendants still run Grandview House, the accommodation inn built by his son. Appreciation for the beauty of this southern lake and its surroundings is widespread, as the government of the day discovered, to its cost, in the 1960s when it proposed a massive raising of the lake to provide more water to feed the West Arm hydroelectricity scheme.

The hydroelectricity potential of the lakes at the bottom of the South Island was recognised as early as 1904 but, except for Monowai, no action was taken until plans took shape for a large aluminium smelter at Tiwai Point near Bluff in the 1960s. As the scheme evolved, and with it the proposal to raise Lake Manapouri substantially, nationwide concern was expressed at the damage such measures would cause to the immediate lake environment. A 'Save Manapouri' campaign hit the headlines in 1969 and remained a major issue until 1972. It resulted in 265,000 people signing a petition opposing the plan, and Norman Kirk's Labour Government came in to power committed to saving the lake. A Bill was passed giving the lake statutory protection, although a modified version of the power scheme went ahead. Today lakes Manapouri, Te Anau and Monowai are watched over by an appointed body of 'Guardians of the Lakes'.

The Manapouri Power Station is located deep underground in the southwestern arm of the lake. Unusually, it does not rely on a high dam to provide water. It takes advantage of the natural 178-metre height difference between Lake Manapouri and the sea at Deep Cove in Doubtful Sound. Penstocks located in tunnels bored into rock beneath the lake deliver water to a machine hall lodged in a massive underground chamber at West Arm, while a tail-race tunnel excavated to Deep Cove in Doubtful Sound takes care of the after-flow. A second tail-race tunnel was completed in 2002 to improve the electricity output from the power station.

Visits to Doubtful Sound or the power station at West Arm depart from Manapouri.

The Waiau River is a favourite destination for those seeking brown trout.

The Oreti River traces a peaceful path through grassy river flats at the end of the Thomson Mountains behind Lake Wakatipu.

Launch trips are available on the lake, the most impressive going from Pearl Harbour on the Waiau River outlet to the head of the lake and the Spey River, where the Doubtful Sound walking track starts.

Waiau River

The Waiau River flows from Lake Te Anau through Lake Manapouri and continues south, traversing both open country and bush-clad gorges before flowing across farmland and into the Tasman Sea at Te Waewae Bay. Once the river with the second-largest flow in New Zealand, it wielded a huge influence on the lives and seasonal patterns of generations of Southland Maori. The river was a major source of food: fern root, eels, shellfish and tutu were gathered in the summer, a range of fish was caught in the autumn, kanakana (lamprey) were caught in the spring, and in winter the people were largely reliant on foods gathered and preserved earlier in the year.

No longer a mighty river, its flow in the section that emerges from Lake Manapouri has been hugely reduced by the water demand of the Manapouri-Deep Cove hydroelectric power scheme.

Both the upper and lower parts of the river provide good fishing and although the Waiau does not have a long estuary as other Southland rivers do, its lagoon is large and well populated with brown trout. Later winter and spring provide the best fishing.

The township of Tuatapere not far from the coast, was originally the site of a punt that ferried passengers across the Waiau River and the town still serves as a gateway into Fiordland National Park.

Mavora Lakes

Not large nor well known outside of Southland, the two Mavora Lakes are a narrow slip of water, joined by the Mararoa River, sitting in a valley at the tail end of the Thomson Mountains behind Lake Wakatipu. We camped there in February one year. Mornings were cool as we waited for the sun to creep down the mountains opposite our campsite beside the upper and bigger of the two lakes. But the days were warm and we enjoyed mountain-biking along the rough track that winds alongside the lake and, in places, through thickets of beech trees.

Early Maori evidently passed this way on a trail from the Oreti River through to the Greenstone River that empties into Lake Wakatipu. Today the Mavora-Greenstone Walkway is a four-day tramping trip through open valley tussockland and beech-forested hill country.

Most people associate the Oreti River with its long, broad estuary that spills out into Foveaux Strait at Invercargill, much enjoyed by anglers. But it starts life as a clear gravelled stream that meanders along a valley south of Lake Wakatipu between the Eyre Mountains and the Mavora Lakes.

The Mataura is New Zealand's most fished brown-trout river. Large and long, it offers hundreds of places that provide excellent trout-fishing. Like all Southland rivers, its trout are difficult to catch – skill and local knowledge are required for success, but Southland anglers are usually willing to provide advice!

The Eglinton River rises on the southeast side of the Main Divide between Lake Wakatipu and Milford Sound and flows south to the eastern shores of Lake Te Anau, following a flat-floored, glaciated valley between beech-clothed mountains. Three glacial lakes – Gunn, Fergus and Lochie, peaty green and immersed in thick bush – sit at the head of the valley, which is a popular scenic reserve for parties en route to and from Milford Sound. Trampers from all over the world visit the area, and anglers go there to try for trout in the lakes and river. A popular walk from the Main Divide is around the north slopes of the Livingstone Range to Key Summit, which commands an extensive view over the Hollyford Valley and Darran Range.

A LAND OF WATER

ABOVE: A shaky walk across the Mararoa River near the Mavora Lakes.

LEFT: The Mavora Lakes in Southland are accessed from Te Anau highway.

THE FAR SOUTH

A LAND OF WATER

LEFT: Contemplating the beauty of Lake Te Anau in late winter.

ABOVE TOP: It takes about one hour to climb to this small mountain tarn at Key Summit, reached from the Milford Road.

ABOVE BOTTOM: Trampers' country — the Caples River in the Humboldt Mountains at the head of Lake Wakatipu.

THE FAR SOUTH

Icy cold and turbulent, the Caples River is typical of the rivers trampers cross when exploring in Fiordland.

Routeburn Falls, on the way up to the Harris Saddle, where the magnificent Falls Hut is located.

Index

Albert Town 122
Alexandra 8, 120, 123, 125
algal blooms 14, 15, 30, 37
Anderson, Mona 100
Aotea canoe 69
Arawa canoe 48
Arthur's Pass 84
Arthur's Pass National Park 87, 100
Aubert, Suzanne 70
Auckland 20
Awatere Valley 96

Balclutha 13, 120, 122, 123
Banks, Joseph 20
Bannockburn 125
Baxter, James K. 69, 71
Bay of Plenty 36–47
Ben Lomond 113
Ben Ohau Range 13
Bethell, Mary Ursula 97
Blenheim 82, 94, 95, 96
boating 19, 37, 38, 65
braided rivers, Canterbury 97–103
Bridge to Nowhere 74
Browning, John 84
Brunner, Thomas 80, 82, 84–86
Bull, James 78
Buller Gorge 88
Bulls 78
bungy jump 113
Butler, Samuel 101

camping 61, 71, 73, 133
canoeing 14, 28, 47, 61, 71 *see also* kayaking
Canterbury 94, 97
Canterbury Plains 97, 100
Caswell Sound 134
Cecil Peak 112
Central North Island, flood waters 75–79
Central Otago 8, 9, 13, 14, 110, 111, 120, 121, 124, 125
Chalmers, Nathaniel 121
Christchurch 15, 100
Cloudy Bay 95
Clyde 123, 124
Clyde Dam 120, 124, 124, 125
coal 83, 85, 86
Colenso, William 59
conservation 14, 102
Cook River Bridge 84
Cook Strait 63
Cook, Captain James 20, 21, 57, 59
Coromandel 21, 24
Coronet Peak ski-field 127
Cotterell, J S 80
Crater Lake (Mount Ruapehu) 52, 53, 54
Cromwell 120, 124, 125
Cromwell Gorge 120, 124

Crown Range 113, 119
Cyclone Bola 57, 60

dams, hydro: 102: Arapuni 34; Aratiatia 30, 31; Atiamuri 30; Aviemore 103, 104; Benmore 104; Clyde 120, 124, 124, 125; Karapiro 30; Maraetai 29, 30; Ohakuri 30; Roxburgh 125; Waipapa 30; Waitaki 104; Whakamaru 26, 30
Dargaville 18, 19
Darran Range 136
Dart River Valley 113
Dawson Falls 65, 67
Dawson, Thomas 65
Deep Cove 135
Dept. of Conservation 30, 47, 59, 81, 94, 95, 105, 134
didymo 81
Dingle Burn 116
Doubtful Sound 135, 136
Dunedin 120, 122, 128, 129, 130
Dusky Sound 84
Dusky Track 133

Earnslaw 112, 113
earthquakes 10, 13, 56, 82, 83, 93
East Cape 46, 56
East Coast (Nth Island) 56–62
East Coast (Sth Island) 94–109
Edgecumbe 45
Egmont National Park 65
Electric Company 123, 124
Endeavour 57
erosion 46, 56, 57, 94, 100
eruptions 49, 52: Mount Tarawera 36, 39, 45; Rotoma 38
Eyre Mountains 136

Falls Hut 141
Far North 18–20
Far South (Southland) 132–41 *see also* Fiordland
Featherston 62
Fiordland 8, 132–135, 140, 144
Fiordland National Park 134, 136
Firth of Thames 20, 21, 26
Firth, Josiah Clifton 22
Fish and Game 15, 95
fishing 14, 20, 23, 28, 30, 37, 39, 46, 47, 51, 57, 58, 61, 63, 65, 69, 78, 84, 85, 86, 96, 102, 105, 113, 125, 133, 136
flood protection 125
floods 5, 13, 23, 44, 45, 47, 60, 65, 75, 76, 77, 82, 83, 95, 100, 112, 121, 125, 129
fossils, shell 60
Foveaux Strait 136
Fox, William 80

Foxton 75, 76
Frankton 110

George Sound 134
geothermal activity 52
geothermal power 49
Gisborne 56, 57, 58, 59
glacial lakes 105, 136
glaciation 132
glaciers 97, 100: Classen 102; Godley 101, 102; Grey 102; Hooker 102, 108; Lyell 101; Maud 102; Ramsay 100; Richardson 102; Rob Roy 110; Tasman 102
Glenorchy 111, 113
Glover, Denis 15
gold 20, 22, 24, 83, 84, 85, 86, 102, 112, 122, 123, 124; dredges 83, 85, 87, 123, 124; panning 84
Great Depression 124
greenstone 84, 86, 87
Greenstone River Valley 111
Greymouth 84, 85, 86
Guthrie-Smith, William Herbert 61

Haast Pass 120
Hackett, A J 113
Hamilton 27, 28, 28, 29, 35
Hanmer Springs 94, 95, 96, 97
Harris Saddle 141
Hastings 57
Hatrick, Alex 71, 72
Hauraki Gulf 19, 20
Hauraki Plains 20–21, 23
Hauraki Plains Drainage Act 23
Hay, Donald 111
Heaphy, Charles 80
Henley 129, 131
Henry Pass 134
Henry, Richard 133, 134
Heretaunga Plains 62
hiking 47, 85, *see also* tramping, walking
Hinekorako 60
Hinemoa 38, 41
Hokianga Harbour 18
Hollyford Valley 136
Horouta canoe 57
hot springs 37, 39, 48, 52, 85
Huiarau Range 46
Huka Falls 26, 33
Humboldt Mountains 139
Hump Track 133
hunting 47, 103, 132, 133
Huntly power station 27, 29
Hutt River Trail 63
hydro lakes 105
hydro lakes, Waikato 30
hydroelectric power 8, 14, 24, 26, 27, 29, 30, 36, 44, 49, 59, 60, 66, 87, 101, 104–105, 124, 130, 133, 135, 142

ice age 100
Idaburn Dam 131
Inangahua 83
Inch Clutha Delta 120
Inland Kaikoura Range 94, 96
Invercargill 136
irrigation 62, 95, 101, 102, 124, 125, 129

Jerusalem 69, 70, 71
jet boat, development of 14
jet-boating 28, 46, 47, 84, 100, 101, 111, 113, 115, 121

Kahurangi National Park 82, 84
Kai Iwi Lakes 18
Kaiangaroa Forest 59
Kaiapoi 100
Kaikoura 96
Kaikouras Seaward 96
Kaimai Ranges 21, 23
Kaimanawa Forest Park 61
Kaimanawa Ranges 77
Kaingaroa Forest 45
Kaipara Harbour 18, 19
Kaipara Heads 19
Kaitawa Power Station 59
Kaitoke Regional Park 63
Karamea 84
Karangahake Gorge 22, 23, 24, 25
Karapiro 30
Karapiro Dam 30
Kawarau 120, 125
Kawarau Gorge 119
Kaweka Forest 61
Kaweka Ranges 62
kayaking 14, 19, 47, 63, 71, 84, 87, 96, 100, 121 *see also* canoeing
Kehu 82
Kepler Track 135
Key Summit 136, 139
Kingitanga 29
Kingston 110, 112
Kinloch 111
Kupe 69

lahar 53
lakes: Alexandrina 108; Aniwhenua 45, 47; Arapuni 30; Aratiatia 30; Atiamuri 30; Aviemore 102; Benmore 102; Benmore 104; Brunner 86–87, 92; Coleridge 101; Constance 81; Cristabel 85; Dunstan 13, 110, 124, 125; Ellesmere 100; Fergus 10, 136, 144; Green (Lake Rotokakahi) 39; Gunn 136; Hankinson 134; Hauroko 132, 133; Hawea 102, 110, 111, 116, 121; Hayes 110; Henry 135; Johnson 115; Kai Iwi 18; Karapiro 28, 29, 30; Lochie 136;

Manapouri 132, 133, 134, 135–36, 142; Manuherikia 110; Maraetai 30, 34; Matheson 15, 91; Mavora lakes 136–37; Middleton 106; Monowai 133, 135; Ohakuri 30; Ohau 13, 97, 102, 105, 106; Okareka 39; Okataina 38, 42; Onoke 62, 63; Poteriteri 132; Pukaki 8, 97, 102, 105, 106; Lake Rotoaira 49; Rotoehu 38, 39; Rotoiti 10, 38, 40, 44, 82; Rotoiti 44; Rotoiti (South Island) 80, 81, 82, 84; Rotokakahi 39; Rotoma 39, 41; Rotomahana 37, 39; Rotorangi 64; Rotoroa 10, 80, 81, 82, 83; Rotoroa (South Island) 80, 88; Rotorua 36, 37, 38, 41, 44; Roxburgh 125, Ruataniwha 105; Sphinx 132; St Patrick 132; Sumner 84; Taharoa 10, 18; Tarawera 37, 38, 39, 40, 41, 42, 45; Taupo 13, 26, 29, 30, 48–52, 54, 69, 78; Te Anau 132, 133–35, 136, 139; Tekapo 97, 102, 104, 105, 108; Tennyson 94, 96; Thomson 134; Tikitapu (Blue Lake) 39, 42; Tutira 15, 61–62; Unknown 10; Waihola 128, 129; Waikapiro 61; Waikare 18; Waikaremoana 10, 58–59, 60; Waikareti 58; Waipapa 30; Waipori 128, 129; Wairarapa 62, 63; Waitaki 102; Wakatipu 8, 110, 111, 112, 113, 115, 116, 136, 139; Wanaka 15, 110, 111, 112, 114, 120, 121; Whakamaru 30; Widgeon 132
Lakes, Kai Iwi 18
Lakes, Mavora 136, 137
Lake Coleridge power scheme 104
Lammerlaw Range 128
landslides 82, 83
Lewis Pass 82, 84, 85
Lewis, Henry 84
Lindis Pass 111
Livingstone Range 136
Lord of the Rings 101
Lower Hutt 63
Lower Taieri Plain 129
Luggate 126
Lyell Creek 83

Mackenzie Basin 104
Mackenzie Country 8, 14
Mackinnon Pass 134
Mackinnon, Quintin 134
Main Divide 84, 86, 100, 136
Main Trunk Line 71, 72, 100
Manapouri Power Station 135

Manapouri-Deep Cove hydro-electric power scheme 136
Manawatu 76
Manawatu Gorge 75, 76
Manawatu Plain 75, 77, 78
Mangaoparo 57
Mangapurua 74
Mangaweka 78, 79
Maniototo 8, 110, 128, 129, 131
Manukau Harbour 20
Maraetai 30
Mararoa 136, 137
Margaret Sievwright Memorial 57
Marlborough 94–96
Martha Mine 24
Marton 78
Maruia Falls 83, 84, 93
Maruia Springs Thermal Resort 85
Matamata 21
Matukituki 110
Matukituki Valley 118
Mavora Lakes 136, 137
Mavora-Greenstone Walkway 136
McMillan, Dr D G 104
Mesopotamia 101
Milford 144
Milford Sound 10, 134, 136
Milford Track 132, 134
miners 112
moa 95, 101, 111, 121, 133
Moana 86
Mohaka Bridge 15, 61
Mokai Gorge 78
Mokoia Island 38, 41
Mokonui Gorge 61
Molesworth Station 94, 95
Morrinsville 21
Motu 47
Motu Challenge 47
Motu Falls 46
Motutaiko Island 49
Mount Algidus Station 100
Mount Aspiring National Park 8, 10, 113
Mount Burke 111
Mount Cook 8, 102, 105, 108
Mount Edgecumbe 36
Mount Maude 111
Mount Ngauruhoe 50, 52
Mount Nicholas 112
Mount Pihanga 69
Mount Roy 111
Mount Ruapehu 26, 49, 50, 52–55 *see also* Crater Lake
Mount Taranaki 13, 54, 64, 66, 67, 69
Mount Tarawera 37, 38, 39 *see also* eruptions
Mount Tongariro 45, 69
Moutere Point 50
Moutoa Island 70
Murchison 82–84
Murchison Mountains 134, 135
Murrell, Robert 135
MV *Cygnet* 66
MV *Ongarua* 70

Napier 15, 57, 60, 62, 76
National Institute of Water & Atmospheric Res (NIWA) 62
National Parks: Arthur's Pass 87, 100; Fiordland 134; Kahurangi 82, 84; Nelson Lakes 80, 82, 84, 94, 96; Te Urewera 46, 58, 60; Whanganui 70, 71, 73; Mount Aspiring 8; Tongariro 26
National Water Conservation Order 47
Nelson 82
Nelson Lakes 80
Nelson Lakes National Park 80, 82, 84, 94, 96
Neumann Range 102
Nevis Valley 121
New Plymouth 66
Ngai Tahu 85, 87, 121, 133
Ngaruawahia 29
Ngatea 21, 23
Ngati Hinehika 60
Ngati Kahungunu 59, 62
Ngati Mamoe 121, 133
Ngati Tumatakokiri 85
Ngati Tuwharetoa 26, 48, 49
Ngatoroirangi 48
Norsewood 75
North Canterbury 94–96
North Otago 111

Oamaru 102
Ohinemuri gold 22, 23
Okere Falls 42, 44
Onepoto Caves 58
Opotiki 46, 47
Orbell, Dr G 134
Owhara Falls 24

Pacific Ocean 13, 96, 100, 102, 121
paddle-steamers 22, 72, 73: *Avon* 28; *Pioneer* 28; *Waipa Delta* 29
Paeroa 21, 22, 23
Palliser Bay 63
Palmerston North 75–77
Panmure 20
Paoa 58
Paparoa Range 86
paragliding 113
Piako River Protection Scheme 23
Pink and White Terraces 37
Piopio 66
Pipiriki 69, 70, 73
pollution 14
Poolburn Dam 129
Port Chalmers 129
Port Molyneux 121
Port Waikato 26
Portland Island lighthouse 60
pounamu 84, 85, 121
Poverty Bay 57
Puhoi 19, 20
Pukaki-Ohau canal 105
Punakaiki 86
Putaruru 21

Queenstown 8, 105, 111–112, 113, 115, 116, 120

rafting 28, 47, 61, 84, 87, 96, 100 *see also* white-water rafting
Raglan 30
Raikaihautu 80
Rakaia Gorge 100
Rakaihautu 13
Ranfurly 110
Rangitata 100
Rangitikei 77
Rautaniwha Plains 62
Read, Gabriel 122
Reefton 83
Rees, William 111, 112
Remarkables 110
Resolution Island 133
Resource Management Act 24
Rimutaka Rail Trail 63
Rimutaka Ranges 63
rivers: Acheron 94, 95; Ahuriri 102–103, 103; Akatarawa 63; Arahura 84, 86, 87; Arnold 86–87; Arrow 111, 112; Awakino 66, 66; Awaroa 18; Awatere (East Cape) 10; Awatere (Marlbrough) 10, 94, 96; Bealey 100; Blue 111; Boyle 85; Buller 81–84, 88; Caples 139, 140; Cardrona 14, 119, 120, 122; Clarence 94, 95, 96; Clinton 134; Clive 62; Clutha 8, 13, 15, 110, 113, 120–27; Clyde 101; Cook 85; D'Urville 81; Dart 111, 115; Dobson 102; Eglinton 136; Esk 100; Fox 85, 86, 93; Glenroy 82, 83; Godley 102, 108; Gowan 80, 81, 82, 83; Greenstone 113, 136; Grey 85, 86; Haast 90; Hakataramea 102; Hangaroa 59, 60; Hangatahua (Stony) 66; Haumoana 62; Havelock 101; Hawea 120, 122; Hikutaia 21; Hooker 102, 108; Hope 82, 85, 96, 96; Hunter 116; Hurunui 84; Hutt 62, 63; Inangahua 83, 85; Kaituna 38, 44; Kapuni 65; Kawarau 110, 113, 119, 121, 123, 124, 126; Kerikeri 18; Koau 121; Lewis 96; Lindis 120; Lowburn 122; Macauley 102; Makarora 120; Makino 61; Manawatu 75–77; Mangakahia 19; Mangamuka 18; Manganui 69; Mangaroa 63; Mangles 82, 83; Manukerikia 120, 122, 124, 125; Maraehara 57; Mararoa 136, 137; Maruia 83, 85; Matakitaki 82, 83, 85; Mataura 110, 121, 136; Mathias 100; Matiri 82; Matukituki 118; Moana 86–87; Mohaka 10, 60–61; Mokau 66, 67; Molyneux 120, 122, 123, 124; Motu 46–47; Nevis 9; Ngaruroro 57, 61, 62; Nokomai 121; Ohau 102; Ohinemuri 22, 23–25; Okuru 111; Ongarue 69; Oreti 136; Otara 46; Owen 82; Pakuratahi 63; Patea 64; Piako 20, 21, 22, 23; Pohangina 76; Pomahaka 120; Poulter 100; Puhoi 19–20; Rainbow 94; Rakaia 97, 100, 101, 104; Rangitaiki 45, 47; Rangitata 97, 100, 101; Rangitikei 77–79; Rees 111; Ruakituri 59, 60; Ruamahanga 63; Sabine 81; Shotover 110, 112, 119, 123, 127; Shotover, Upper 123; Spey 135, 136; Stony 66; Taharua 61; Taieri 122, 128–30; Tamaki 20; Tangarakau 69; Taramakau 87; Tarawera 36, 45; Taruheru 57; Tasman 102; Tauherenikau 62; Tekapo 102; Tokaanu 48; Tongariro 26, 51; Travers 80; Tuapeka 120; Tukituki 57, 62; Turanganui 57; Tutaekuri 57, 62; Upper Grey 85, 86; Upper Mohaka 60; Upper Shotover 123; Upukerora 123; Waiau 56, 57; Waiau (Marlborough) 96, 97; Waiau (Southland) 133, 134, 135, 136; Waihou (Hauraki) 20, 21, 22, 23; Waihou (Northland) 18; Waikaretaheke 58, 59; Waikato 26–35, 49, 52; Waimakariri 97, 100; Waimata 57; Waioeka 46, 47; Waipaoa 58; Waipawa 62; Waipori 8, 129, 130; Waipunga 61; Wairau (Marlborough) 10, 94–96; Wairaurahiri 133; Wairoa (East Coast) 19, 44, 59–60; Wairoa (Northland) 18–19; Wairua 19; Waitaki 14, 101–102, 104; Waitara 64; Waitawheta 23; Waitoa 21, 23; Wanganui 84, 88; Whakapapa 15, 69; Whakatikei 63; Whangaehu 53; Whangamomona 69; Whanganui 15, 30, 64, 69–74; Whataroa 84; Wilberforce 100; Wilkin 111; River Road 71, 73
riverboats 28, 72
Roaring Meg Power Station 121
Rock and Pillar Range 128
Rotoiti Nature Recovery Project 81
Rotoma eruption 38
Rotoroa 10, 15, 21, 28, 36, 37, 58, 81
Rotorua lakes 38–43
Routeburn 15
Routeburn Falls 141
rowing 28
Roxburgh hydro dam 124
Royal Forest and Bird Protection Society 95
Ruahihi Power Station 44
Ruahine Range 62, 75, 76, 78
Ruamano 60
Ruatoria 56, 57

sailing 28
Save Manapouri campaign 135
schist 132
Scotts Ferry 77
Seaward Kaikouras 96
seiche 111
shell fossils 60
ship building 19
shipping 73
skifields 52
Skippers Canyon 112, 127
South Otago 120
Southern Alps 84, 97, 100, 120, 132
spa towns 22, 36, 37
Spenser Mountains 82, 94, 96
Sphinx Lake 132
St Arnaud 80, 81, 94
St Arnaud Range 80
St Bathans 110
St James Station 96
steamer boats 22, 112
Stone Store 18
Strath Taieri 128
surf-casting 47
Sutherland Falls 132
swimming 39, 47, 58, 66

Taieri Ferry, Upper 129
Taieri Gorge 128, 130
Taieri Plain 8, 128, 129, 131
Taieri River Mouth 128, 129
Taihape 78
Tainui 27
takahe 134, 135
Takitimu canoe 59, 133
Takitimu Marae 59
Tama Lakes 52
Tamaki 20
Tamatea 133

Tangiwai 53
Tangiwai Memorial 52
Tapuaeroa 57
Taranaki 64–67
Tararua Range 63, 75
Tarawera Falls 45
Tasman Sea 18, 20, 64, 75, 82, 84, 86, 136
Taumarunui 69, 74
Taupo 26, 28, 48
Taylor, Richard 73
Te Ana-Au Caves 134
Te Anau 133
Te Araroa 56
Te Arawa 45
Te Aroha 21, 22
Te Heuheu, Tukino 26
Te Hoe Gorge 61
Te Kooti 47, 49
Te Kuiti 66
Te Marua 63
Te Papanui Conservation Park 128
Te Rakaihautu 110
Te Rauparaha 111
Te Reinga Falls 60, 61
Te Urewera National Park 46, 58, 60
Te Waewae Bay 133, 136
Te Wairoa 37, 59
Tekapo 105
Tekapo power station 105; Tekapoa A 104; Tekapo B 105

Thomson Mountains 136
Timaru 102
timber milling 83, 112
Tiwai Point 135
Tokaanu 52
Tokaanu Power Station 48
Tongariro National Park 26
Tongariro Power Devel 30
tourist centre 49
tramping 58, 71, 103, 132, 133, 136 see also hiking, walking
transport 19, 20, 22, 28, 29, 66, 71, 76, 112, 129
Treaty of Waitangi 27, 28, 70
Tuai Power Station 59
Tuapeka 122
Tuhoe 58
Tutanekai 38, 41
Twizel 104, 105
Upper Hutt 63
Urewera Range 58
Uruao 13

van Asch, Henry 113
vineyards 113
volcanoes 36, 50, 52; eruptions 13, 36, 48, 80

Waiau 59
Waiau Ferry Bridge 96
Waihi 24, 52
Waihi Gold Mining Company 29

Waihou Valley Scheme 23
Waikaremoana 59
Waikaremoana Great Walk 58
Waikato 28
Waikato peat lakes 30
Waimarie paddle-steamer 73
Waioeka Gorge 47
Waioeka Scenic Reserve 47
Waipukurau 62
Wairarapa waterways 62–63
Wairau Plain 95
Wairoa 58, 59, 60, 61, 62
Waitahuna 120
Waitaki 13, 14, 97, 104
Waitaki Dam 104
Waitaki Power Scheme 104, 105
Waitara 66
Waitara Bridge 65
Waitomo 28
walking 18, 39, 57, 65, 66, 71 see also hiking, tramping
Walter Peak 112
Wanaka 105, 113, 120, 125,
Wanganui 69, 70, 72, 73
Wanganui Bluff 88
War Birds Over Wanaka 113
Water Conservation Order 61, 101
water sports 30, 39 see also rafting, rowing, sailing, water-skiing, white-water rafting, wind-surfing
water treatment 24

water-skiing 28, 39
Wellington waterways 62–63
Wenderholm 20
West Arm Hydroelectricty scheme 135
West Coast (South Island)10, 80–93 111, 126
Westport 82, 83, 84
wetlands 30, 63, 96, 103
Whakamaru Dam 26
Whakarewarewa 37
Whakariki Point 57
whaling 59, 69
Whangamata 23
Whanganui National Park 70, 71, 73, 74
Whanganui 14
Whanganui River Trust 72
Whangape Harbour 18
Wheao Hydroelectric Scheme 45
whitebaiting 65, 66, 83, 86
white-water rafting 44, 47, 65, 86 see also rafting
Wi Pere Memorial 57
Wiffen, Joan 61
wildlife 96
Wilmot Pass 135
wind-surfing 28, 125
Woodville 75, 76
World Wetlands Day 30

Young Nick's Head 58
Young, Nicholas 57

Bibliography

Angus, John H. *Aspiring Settlers*, John McIndoe, Dunedin, 1981
Dixon, Maren and Ngaire Watson eds, *A History of Rangiotu*, Dunmore Press Ltd, Palmerston North, 1983
Ell, Sarah ed. *Pioneer Women in New Zealand – from their letters, diaries and reminiscences*, The Bush Press, 1992
Forest & Bird, *Royal Forest and Bird Protection Society Magazine*
Hall-Jones, John. *Fiordland Explored*, AH & AW Reed, Wellington, 1976.
Lind, C A. *The 100 Year Flood*, Craig Printing Co Ltd, Invercargill, 1978
McKenzie, Alice ed. *Pioneers of Martin Bay*, Whitcombe and Tombs Ltd, 3rd edn, 1970
McLintock, A H, *The History of Otago*, Otago Centennial Historic Publications, 1949.
Miller, F W G. *Golden Days of Lake Country*, Otago Centennial Historical Publications, Dunedin, 1949
Moon, Geoff and Sue Miles. *The River: The story of the Waikato*, Heinemann, Auckland. 1984
Moore, C W S. *The Dunstan: A history of the Alexandra–Clyde districts*, Otago Centennial Historical Publications, 1953, Capper Press, Christchurch, 1978.
More, Vennell and David More. *Land of the Three Rivers*, Wilson & Horton, Auckland. 1976
Peat, Neville and Brian Patrick. *Wild Central*, University of Otago Press, Dunedin, 1999
Peat, Neville and Brian Patrick. *Wild Rivers*, University of Otago Press, Dunedin, 2001
Purdie, Margaret ed. *Skirt Tales*, Pope Print, Timaru, 1995

Roxburgh, Irvine. *Wanaka Story*, Otago Centennial Historical Publications, 1957.
Tyrrell, A R. *River Punts and Ferries of Southern New Zealand*, Otago Heritage Books, 1996
Vercoe, Graham. *Bow Waves on the Waikato*, Reed Books, Auckland. 1997
Woodhouse, A E. *Tales of Pioneer Women*, Silver Fern Books Ltd, Hamilton, 1988
Young, David and Bruce Foster. *Faces of the River*, TVNZ Publishing, Auckland, 1986
Young, David. *Woven by Water: Histories from the Whanganui River*, Huia Publishers, 1998

Websites

Department of Conservation (DOC), www.doc.govt.nz
Dictionary of New Zealand Biography, www.dnzb.govt.nz
Encyclopaedia of New Zealand, www.teara.govt.nz
Fish and Game NZ, www.fishandgame.org.nz
Ministry for the Environment, www.mfe.govt.nz/publications – *State of New Zealand's Environment*, 1997

Photography credits

All photographs taken by Russell McGeorge except:
Horizons' Regional Council p 77; IGNS p 62, p 63